AF486777

STRATEGIC SAFETY

Charting a Course for Safety Excellence

Don Fritz, CSP

Sanovus

Acknowledgements

I am deeply grateful to my wife, Jennifer, who is also my personal grammarian, for her support throughout the development of this book. Not only was she vey supportive during the countless hours I worked on it, but she also spent hours herself helping with the editing of it. Her editing contribution significantly improved the quality and flow of the book's content.

In addition, I would like to thank the numerous supervisors, colleagues and co-workers who contributed to the successes I have enjoyed over the years in applying some of the concepts presented. Strategy execution is a team sport, and I was fortunate to be a part of winning teams.

Lastly, this book would not have been possible without the knowledge of the cited academics, business leaders, and safety professionals who are the true experts on the various topics covered. I am indebted to them for their contributions to the discipline of strategic safety espoused by this book.

Table of Contents

List of Tables

List of Figures

Introduction

Are you looking for a better direction to take your safety program in, but aren't sure which way to go or which path to take? If so, hopefully you don't find yourself in the same mindset Alice was in while taking her journey through the mad, illogical, and nonsensical world of Wonderland.

At one point, Alice happened upon the Cheshire Cat sitting in a tree before her. She decided to ask the cat for advice on which way to go.

> 'Cheshire Puss,' she began, rather timidly, as she did not at all know whether it would like the name: however, it only grinned a little wider. 'Come, it's pleased so far,' thought Alice, and she went on. 'Would you tell me, please, which way I ought to go from here?'
>
> 'That depends a good deal on where you want to get to,' said the Cat.
>
> 'I don't much care where—' said Alice. '
>
> Then it doesn't matter which way you go,' said the Cat.
>
> '—so long as I get somewhere,' Alice added as an explanation.
>
> 'Oh, you're sure to do that,' said the Cat, 'if you only walk long enough.'[1]

Alice, unfortunately, did not have a clear idea of where she wanted to go. Steve Maraboli, inspirational speaker and author, says in his book *Life, the Truth, and Being Free*, "A lack of clarity could put the brakes on any journey to success."[2] To succeed in improving a safety program, you need a clear path forward.

This book aims to help you chart a course for safety excellence. What is safety excellence? Is it zero injuries? Should we define it in terms of performance, such as using injury rates, or something else? In an article titled "Rethinking Safety Excellence," published in *Occupational Health & Safety*, Shawn M. Galloway, now CEO of ProAct Safety, stated:

> What we have defined as great in safety has been largely and historically determined by incident and injury rates and only surface-level and subjective measurements about the organization and its culture. All progress begins by thinking differently. It's time we rethink what constitutes excellence in safety.[3]

Galloway then described a better definition of safety excellence using three characteristics of organizations that achieve it:

1. The Ability to Achieve and Repeat Great Results

2. Keen Insight into What Led to the Results

3. A Cultural Mindset of Continuous Improvement[4]

Excellence is not perfection; it is excelling at something, being extremely good at it. Is your organization excelling in safety? Excellence is not a one-time achievement. As Galloway suggests, excelling at safety means having exceptional results year after year, and knowing why you achieved and maintained those results. The question is, "How can you get to that point?" The answer is strategy.

My appreciation for strategy began when I was a young boy, an appreciation that stuck with me throughout my career as a Certified Safety Professional. My brother and I were born just a year apart, so we enjoyed competing against each other throughout our childhood, especially playing board games. One of our favorites was *Stratego*. The game's name is based on the Byzantine Greek word *strategos*, meaning *general*. The modern term *strategy* comes from that word.

The military-based game required us to employ various strategies to capture our opponent's flag first. While simple enough for children to play, *Stratego* is one of the most complex board games, even more complex than chess in certain regards.

My appreciation for strategy grew when I enlisted in the U.S. Navy, and played a small part in our nation's military strategy during the 1980s. I spent the first two years of my six-year enlistment attending various training schools, including the prestigious Navy Nuclear Power School. Upon graduation, I spent two years as a junior staff instructor at that school. During the last two years of my enlistment, I was attached to the nuclear-powered ballistic missile submarine *USS George C. Marshall* (SSBN-654). It was one of the "41 For Freedom," submarines designed to be a strategic deterrent against the threat of nuclear war. That Navy experience furthered my appreciation for strategy, real military strategy.

Upon leaving the Navy, I began a career in the power generation industry. I spent over twenty-five years managing Environmental Health and Safety (EH&S) at the business unit, regional, divisional, and corporate levels. Throughout my career, I placed a strong emphasis on my professional development. One of the focal points of my study was strategy.

Later in my career, I had the privilege of serving as a Director of EH&S for a medium-sized, international, privately owned joint venture. The two companies involved were independent service providers for the power generation, oil and gas, industrial, and aerospace industries. I had divisional and corporate safety oversight of over thirty power plants, sixteen manufacturing and repair shops, and eight field service operations.

My primary purpose from the start was to combine the safety programs and cultures of the two companies. The objective was to drive continual improvement in safety performance in the process. In our industry, injury rates were a critical factor in our ability to bid on projects and win long-term service agreements. This made safety performance critical to the joint venture's success.

Our initial efforts involved taking a data-driven strategic approach to injury reduction. Our broader safety strategy included three to five strategic objectives that drove targeted safety improvement programs, campaigns, and initiatives down through the organization's business units around the world.

These business units developed their own annual improvement plans that adopted and aligned with the corporate strategy, and addressed local safety issues. We had over fifty full-time safety staff throughout the organization, and just as many operations managers, involved in our annual planning process.

After five years, our combined efforts achieved a 77% reduction in Total Recordable Injury Rate (TRIR), resulting in a TRIR 80% below our weighted industry average. Our strategy was the driving force behind those results, along with the strong commitment to and passion for safety exhibited by those involved in executing it.

While completing our annual planning process, I noticed a general lack of knowledge and skills in strategic thinking and planning in our organization, myself included. This was apparent in the business units' annual plans, which took a brute-force approach to improving safety. Unfortunately, I could not find suitable resources that taught safety strategy. This book is my way of closing that knowledge gap within the safety profession.

This book does not approach the concept of strategy from an academic perspective, though it contains many references from experts on various aspects of the subject, all cited for further study. Instead, the book gives you, as a safety professional, a practical approach to strategic thinking, analysis, and planning. It takes what experts in the various fields of study have learned and translates it into practical knowledge and resources.

In this book, you will learn principles and techniques to help you take a strategic approach to safety. This includes a three-phase strategic thinking process that applies to any safety challenge you face. It also includes several strategic analysis methods for the four Strategic Areas of Safety. You will learn a method for selecting strategic objectives, for a safety strategy, that is comprehensive, systematic, and objective. Finally, this book will teach you how to bridge the infamous strategy execution gap.

In short, this book will give you clarity for a successful path forward. You won't have to be like Alice, asking directions with nowhere in mind to go. You will know exactly where to go and how to get there.

Don Fritz, CSP

Endnotes

1. Lewis Carroll. *Alices's Adventures in Wonderland*, (Chicago: VolumeOne Publishing, 1998), 89-90.

2. Steve Maraboli, *Life, the Truth, and Being Free*, (Port Washington: A Better Today Publishing, 2014), 38.

3. Shawn M. Galloway, "Rethinking Safety Excellence," *Occupational Health & Safety*, October 1, 2016, accessed October 13, 2023, https://ohsonline.com/articles/2016/10/01/rethinking-safety-excellence.aspx

4. Ibid.

Artificial Intelligence

Before getting to the content of the book, I should say a word or two about Artificial Intelligence (AI), given its rising popularity at the time of publishing. First, AI was not used to write this book, but did play a part in some of the editing of it.

Second, the use of AI did not become popular until the very end of my full-time professional career. Therefore, I did not use AI during my career. As a result, AI was not included in the content of this book. AI is, however, being used more and more in the safety profession, especially in safety management software.

As a professional member of the American Society of Safety Professionals (ASSP), I appreciate the sensible approach they are taking to assist safety professionals in using AI in a responsible, ethical manner. They have "formed a task force that will explore the role of AI in the future of work and safety. Their goal will be to better understand how AI impacts the workplace, the workforce, members and the Society's role as a trusted source."[1] They state the following about AI:

> Like any emerging/new technology, artificial intelligence (AI) is a tool that has the potential to improve occupational safety and health (OSH) and promote healthier work environments that enhance worker well-being overall. At the same time, it's critical to recognize that AI also has the potential to introduce new hazards that endanger workers, as well as new concerns about privacy.[2]

Setting the employee privacy issue aside, the concern I have regarding the use of AI in safety is its potential for introducing unintended risk. Not only would introducing new hazards raise the risk to employees, it could also present legal risks for safety professionals.

Fisher Phillips is a large international law firm providing services in AI, data, and analytics, as well as workplace safety, among other services. They speak of the legal risk on their website.

> One of the most significant legal risks associated with AI in workplace safety is determining liability when AI systems are involved in safety-related decisions. If an AI system fails to identify a hazard, makes an incorrect prediction, or generates inaccurate recommendations, the consequences could be severe – including accidents, injuries, or fatalities. This raises important questions about accountability and liability.[3]

Perhaps the best way to mitigate this risk is to have a good foundation of safety knowledge from which to evaluate AI generated solutions.

The question, though, as it relates to the topic of this book, is: "How does AI fit in with a company's plans to develop a safety strategy?" Let us ask a strategy expert.

Adam Asch is a Senior Consulting Associate with The Balanced Scorecard Institute (BSI), a Strategy Management Group company. He states:

Artificial Intelligence (AI) is revolutionizing how organizations approach strategy and strategy execution. From predictive analytics to process automation, AI offers transformative capabilities that enable businesses to make data-driven decisions with unprecedented speed and accuracy. However, while AI can significantly augment strategic processes, overreliance on these systems without human oversight introduces critical risks. The right balance between AI's capabilities and human judgment is essential for sustainable success.[4]

While AI can significantly enhance strategic planning, such as with predictive safety analytics, human knowledge and judgment is still needed. There is something known as the 30% rule of AI. It suggests that about 30% of tasks in complex roles can and should be automated with AI, whereas the other 70% requires human expertise and oversight.

The bottom line, when it comes to strategy, is that AI cannot formulate strategies, it can only augment the strategy development process.

This book aims to help you develop a better understanding of strategic safety. If you do employ the use of AI in your strategy development, you will be able to provide the human oversight needed to ensure your AI supported strategy makes sense.

Endnotes

1. "The Role of AI in the Future of Work and Safety," *American Society of Safety Professionals*, accessed March 10, 2026, https://www.assp.org/about/artificial-intelligence---safety.

2. Ibid.

3. "Artificial Intelligence for the Safety Professional – Benefits, Risks, and Legal Implications," Fisher Phillips, March 14, 2025, accessed March 10, 2026, https://www.fisherphillips.com/en/insights/insights/artificial-intelligence-for-the-safety-professional-benefits-risks-legal-implications

4. Adam Asch, "(Chicagougmented Strategy: The Promise and Pitfalls of AI in Strategic Planning," *Balanced Scorecard Institute*, Feb 24, 2025 accessed March 10, 2026, https://balancedscorecard.org/blog/augmented-strategy-the-promise-and-pitfalls-of-ai-in-strategic-planning/

PART I
Strategic Safety

Chapter One:
Strategy

Most organizations have safety wishes or goals; few have true strategies. — Terry L. Matthews & Shawn M. Galloway

What is the first thing that comes to mind when you think of strategy? Does safety come to mind? Probably not. For many, strategy is associated with military strategy, and rightly so. For others, it is associated with business strategy.

Before delving into the weightier topics of strategy, it will help if we take a look at the meaning of the term itself. We will then learn how crucial strategy is to achieving victory in war or success in business, and how it can help you chart a course for excellence in safety.

Rich Horwath, CEO of the Strategic Thinking Institute, provides an in-depth etymology of *strategy* in his paper titled "The Origin of Strategy."

The term "strategy" is derived indirectly from the Classic and Byzantine (330 A.D.) Greek "strategos," which means "general." While the term is credited to the Greeks, no Greek ever used the word. The Greek equivalent for the modern word "strategy" would have been "strategike episteme" or (general's knowledge) "strategon sophia" (general's wisdom). One of the most famous Latin works in the area of military strategy is written by Frontius and has the Greek title of Strategemata. Strategemata describes a compilation of strategema, or "strategems," which are literally "tricks of war." The Roman historians also introduced the term "strategia" to refer to territories under control of a strategus, a military commander in ancient Athens and a member of the Council of War

The word strategy retained this narrow, geographic meaning until Count Guibert, a French military thinker, introduced the term "La Strategique" in 1799, in the sense that is understood today. Consequently, neither the military community before Count Guibert nor the business community before H. Igor Ansoff (Corporate Strategy, 1965) could see the strategic element in their domains clearly enough to give it a name.[1]

Based on the term's origin, strategy refers to generals applying their knowledge to deploy tricks of war to win victory over their enemy. You may recall the mythological trick of war the Greeks deployed against the Trojans to gain entrance into Troy during the Trojan War. That trick was the Trojan horse, and, according to legend, it proved to be a winning strategy.

The online Cambridge Dictionary defines *strategy* as: "A detailed plan for achieving success in situations such as war, politics, business, industry, or sport, or the skill of planning for such situations."[2] When we use the term, we are not referring to some nebulous concept only academic scholars and business executives can understand; we are simply talking about a detailed plan for success, or the skill of developing that plan.

Military Strategy

For millennia, the discipline of strategy has been a critical field of study for those providing leadership in war. The earliest and most influential text written on strategy in warfare, written long before the term strategy existed, is *The Art of War*. Though some mystery still surrounds his identity, Chinese general Sun Tzŭ (544-496 B.C.) is credited with writing it.

In 1910, Lionel Giles wrote the most notable English translation, *Sun Tzŭ on the Art of War*. Thanks largely to that work, Sun Tzŭ's ancient classic remains an essential study for military strategists today.

The Art of War was written during China's Spring and Autumn period of the Eastern Zhou dynasty, which marked the erosion of the dynasty's royal power as the feudal states exercised increasing political autonomy. Those states started forming alliances and waging wars with other feudal states.

The Art of War was written to help feudal states wage wars more strategically; wars that were shorter, less costly, and less brutal. The ultimate purpose was to help put an end to the constant wars. While some of Sun Tzŭ's teachings would not apply to modern warfare, many of its principles are timeless.

Sun Tzŭ placed the highest priority on strategy. He wrote: "Thus it is that in war the victorious strategist only seeks battle after the victory has been won, whereas he who is destined to defeat first fights and afterwards looks for victory."[3] Ho Shih, one of the many commentators on *The Art of War*, explains this statement. He wrote, "In warfare, first lay plans which will ensure victory, and then lead your army to battle; if you will not begin with stratagem but rely on brute strength alone, victory will no longer be assured."[4]

Sun Tzŭ also wrote: "All men can see the tactics whereby I conquer, but what none can see is the strategy out of which victory is evolved.[5] While tactics might win battles, strategy wins the war. As safety professionals, we would be well-served to apply some of the teachings in *The Art of War*. It will get you thinking more strategically about your own safety challenges, and help you resolve them more efficiently and effectively.

Business Strategy

Business professionals began applying the concept of strategy in the mid-20th century. Academics and business leaders started realizing that the principles involved in gaining victory over enemies could apply to gaining an advantage over competitors and taking their market share.

In 1962, Harvard Business School professor and academic researcher Alfred Chandler Jr. wrote a book titled *Strategy & Structure*. In that book, Chandler defined strategy from a business perspective: "Strategy can be defined as the determination of the long-term goals and objectives of an enterprise, and the adoption of courses of action and the allocation of resources necessary for carrying out these goals."[6]

In 1965, Russian-American mathematician and business manager Igor Ansoff wrote his landmark book *Corporate Strategy: An Analytic Approach to Business Policy for Growth and Expansion*. That book established strategy as an individual field of study within business. It was Ansoff, known as the father of strategic management, who was credited with coining the terms *strategic planning* and *strategic management*.

In 1985, Harvard Business School professor Michael Porter wrote his second influential book on strategy, *Competitive Advantage: Creating and Sustaining Superior Performance*. Porter is known as the father of the MBA (Master of Business Administration) and strategy design. The practice of using strategy to gain a competitive advantage over business competitors has since become a critical responsibility of corporate executives.

ClearPoint Strategy markets strategic planning software that brings together all the tools needed to execute a strategic plan. They provide an article on their website titled "Why Is Strategic Planning Important? & 4 Benefits." That article lists four reasons strategic planning is crucial to achieving success in business.

1. *Direction:* Strategic planning offers a sense of direction and outlines measurable goals. It's a tool that's useful for guiding day-to-day decisions and also for evaluating progress and changing approaches when moving forward.

2. *Future Focus:* Strategic planning allows organizations to anticipate and respond to changes in the business environment. It also helps to forecast potential opportunities and threats, which is crucial for surviving in today's dynamic business world.

3. *Operational Efficiency:* Strategic planning provides the basis for all management decisions, reducing the potential for wasted resources, missteps, and inefficiencies.

4. *Competitive Advantage:* A strategic plan allows organizations to foresee their future and to prepare accordingly. Organizations that plan strategically are better equipped to predict the market, anticipate changes, understand competitors, and make decisions that keep them ahead.[7]

Another article on ClearPoint's website presents over fifty eye-opening strategic planning statistics. The final one states. "Organizations that are able to successfully unlock the capacity to execute new growth strategies increase their profitability by 77%."[8] That statistic shows how beneficial strategy can be to the success of a business. A proverb says, "The proof of the pudding is in the eating." A 77% increase in profitability would be delicious pudding for any CEO, as would a 77% reduction in injuries to a safety professional.

One example of a successful growth strategy, and applying principles from the *Art of War*, is the growth of Netflix and other streaming providers. Netflix went from being a DVD-by-mail rental service to being the most profitable standalone streaming service.

Sun Tzŭ wrote: "If you know the enemy and know yourself, you need not fear the result of a hundred battles."[9] Streaming services knew the disadvantages of the broadcast and cable television business, and how many viewers preferred lower cost subscriptions and on-demand viewing, myself included. The high monthly costs and rigid

TV schedules displeased many cable viewers. As a result, between the cord-cutters and the binge watchers, Netflix and other streaming services captured a significant share of the TV viewership market over the years. Advances in technology played a big part, not the least of which was the advent of smart TVs, as did the pandemic lockdowns.

By 2025, streaming held approximately a 47.5% share of total U.S. TV viewing, while broadcast TV dropped to 21.4% and cable dropped to 20.2%. Netflix has the largest streaming viewership, with over 325 million subscribers, holding a 20% share of the competitive streaming market.

Netflix began their streaming service in 2007. Their strategy of transitioning the business from rental DVDs to a streaming service certainly was a winning strategy. Prior to that transition, Netflix only had a little over six million subscribers for their DVD rentals.

Safety Strategy

You may ask how the success of military or business strategy relates to safety strategy? That is a valid question, especially considering what Mr. Strategy says on the subject.

Kenichi Ohmae is a Japanese organizational theorist and management consultant widely known as Mr. Strategy. He introduced the Japanese management and strategic planning methods to the Western world. In his seminal book *The Mind of the Strategist: The Art of Japanese Business*, Ohmae writes:

> Without competitors there would be no need for strategy, for the sole purpose of strategic planning is to enable the company to gain, as efficiently as possible, a sustainable edge over its competitors. Corporate strategy, thus, implies an attempt to alter a company's strength relative to that of its competitors in the most efficient way.[10]

So that settles it, right? Mr. Strategy says strategy is not needed if you have no competitors, or enemies in the case of war. Since the safety function has no competitors or enemies, we do not need strategy, right? Wrong! Consider the following:

Safety Has Competitors

Your organization does not have a monopoly on the goods or services it provides. It has competitors, and to an increasing extent, its potential customers are including social responsibility factors such as safety performance in their selection of providers. This is especially true of some service providers where low injury rates are required to bid on projects or service contracts. Here, negative safety performance can be detrimental to your company's sales.

Safety Has Enemies

While safety professionals are not at war with literal enemies, we are at war with figurative enemies. These enemies are hazards, and the causal factors that weaken your defenses against their harmful effects. With weakened defenses, your risk of experiencing an event such as an injury, process safety failure, vehicle accident, fire or explosion, or environmental release increases. You can take the reactive Whac-a-Mole approach to preventing future incidents like these and only address causal factors when they surface. Or, you can take a proactive, strategic approach to ensure victory.

Given the positive impact a strategy can have on achieving an organization's safety goals and objectives, you would think that more organizations would develop a safety strategy. Many think they have a safety strategy, but do they really have one?

ProAct Safety is a leading safety consulting firm. Terry L. Mathis, its founder (retired), and Shawn M. Galloway, its current CEO, coauthored a book titled *STEPS to Safety Culture Excellence*. STEPS stands for Strategic Targets for Excellent Performance in Safety. In it they state: "The best tool for leadership/management to impact safety culture is the development of a safety strategy. Most organizations have safety wishes or goals; few have true strategies."[11]

Consider the strategy I described in the introduction. We began with an attribute analysis to target our most prevalent injury types. We then created comprehensive plans to prevent their recurrence. Compare that to the reactive Whac-a-Mole approach, addressing individual injuries when they surface, which some might incorrectly consider a strategy. Which one do you think offered the greater assurance of success?

ProAct Safety includes strategy as one of its guiding principles.

Many organizations do not have overarching safety strategies; they have goals and programs to meet them. ProAct Safety helps organizations develop the kind of safety strategies that enable true excellence in performance that are repeatable year after year.[12]

They get it! Real strategies produce real results, and developing an overarching safety strategy will lead to sustainable success.

I trust this chapter portrayed how crucial strategy is to achieving victory in war or business, and how it can help you have success in safety. The next chapter explores taking a more strategic approach to safety.

Endnotes

1.   Rich Horwath, "The Origin of Strategy," *Strategic Thinking Institute*, accessed October 13, 2023, https://www.strategyskills.com/the-origin-of-strategy/.

2.   Cambridge Dictionary, s.v. "Strategy," accessed October 15, 2023, https://dictionary.cambridge.org/dictionary/english/strategy.

3.   Lionel Giles, *Sun Tzŭ on the Art of War*. [1910; Project Gutenberg, 2004], Chapter 4, Paragraph 15, accessed October 15, 2023, https://www.gutenberg.org/files/132/132-h/132-h.htm

4.   Ibid.

5.   Ibid., Chapter 6, Paragraph 27.

6.   Alfred D. Chandler, Jr., Strategy and Structure: Chapters in the History of the American Industrial Enterprise (Cambridge: The MIT Press,1962), 13.

7.   Laurel Miyake, " Why is Strategic Planning Important? & 4 Benefits " *ClearPoint Strategy,* accessed October 15, 2023, https://www.clearpointstrategy.com/blog/the-benefits-of-strategic-planning.

8.   Ted Jackson, "50+ Eye-Opening Strategic Planning Statistics [2022]," *ClearPoint Strategy*, accessed October 14, 2023, https://www.clearpointstrategy.com/blog/strategic-planning-statistics.

9.   Lionel Giles, *Sun Tzŭ on the Art of War*. [1910; Project Gutenberg, 2004], Chapter IIII, Paragraph 18, https://www.gutenberg.org/files/132/132-h/132-h.htm.

10. Kenichi Ohmae, *The Mind of the Strategist: The Art of Japanese Business* (New York: McGraw-Hill,1982), 36.

11. Terry L. Mathis and Shawn M. Galloway, *STEPS to Safety Culture Excellence* (Hoboken: John Wiley & Sons, Inc., 2013), xxiii.

12. "About," *ProAct Safety*, accessed October 14, 2023, https://proactsafety.com/about.

Chapter Two:
Strategic Safety

The discipline of strategy has such great potential for helping you achieve your safety goals and objectives, that it warrants making it an integral part of your approach to safety. In fact, it would benefit the safety profession if strategy became its own field of study within safety management, as it did in business management. Dare we call it "Strategic Safety?"

One of the beauties of strategic safety is that you do not need to dismiss anything you have learned so far about safety. All of that knowledge and experience is still needed. The change comes in knowing when and where to apply that knowledge and experience. Strategic safety is essentially a new approach to safety, one that helps you work smarter, no harder.

When the business community first introduced the discipline of strategy, the discussion centered on how strategic management differed from traditional management. Strategic safety warrants the same. The aim here is to define the term and explain how it differs from traditional safety. In doing so, we will start with what strategic safety is not, and then go on to explain how it provides a distinctive approach to safety.

21

Defining Strategic Safety

What It Is Not

Before defining *what* strategic safety is, it is important to describe what it *is not*. It is not a new school of thought in safety. There are three primary schools of thought on how best to approach safety.

1. *Systems-based Safety:* This school of thought believes it is failures of safety management systems or gaps in incident prevention barriers that cause injuries. It emphasizes improving these systems and closing the gaps to prevent injuries. It applies the principles of systems theory.

2. *Behavior-based Safety:* This school of thought believes that unsafe behavior causes injuries. It emphasizes changing employee behavior to prevent injuries. It applies the principles of behavioral psychology.

3. *Culture-based Safety:* This school of thought believes that the ultimate cause of bad safety performance is a poor safety culture. It emphasizes developing a positive, proactive, and participative safety culture to improve safety performance. It applies the principles of organizational psychology.

Each of these schools of thought has its own merits, and each provides useful tools in ensuring a safe workplace. Strategic safety, though, does not espouse a unique set of opinions regarding safety, which is what a school of thought does. Rather, it applies a deeper and higher level of thought to safety. Strategic safety is an approach to safety that uses the principles of strategy to drive continual improvement in safety performance.

Define Your Terms

In his encyclopedic dictionary *Dictionnaire philosophique portatif* (1764), Voltaire wrote: "Define your terms, you will permit me again to say, or we shall never understand one another."[1] Heeding Voltaire's warning, let us examine what strategic safety is by defining the component terms.

From a grammatical perspective, the term *strategic safety* contains an adjective and a noun. An adjective modifies or describes a noun, and in this case, *strategic* is modifying or describing *safety*.

Safety

The word *safety* is used here in the same way it is used throughout the safety profession, referencing all aspects of occupational safety. When a company states: "Safety is our top priority," it is generally referring to the well-being of people, plant, and property, and all that the company does to ensure it. When you think of strategic safety, then, think of *safety* in how the term is used throughout the safety profession.

Strategic

The operative word in the term *strategic safety* is *strategic.* The primary definition is something that relates to, or is characteristic of, strategy. *Strategy* can be defined as detailed or long-range plans for achieving something or reaching a goal, or the skill of making such plans. Strategic safety, in the broader sense, can be defined as safety that is strategy-driven, safety that uses strategic thinking, analysis, and planning.

One way to understand strategic safety that is strategy-driven is to compare it to safety that is compliance-driven. Some organizations focus only on complying with OSHA regulations or industry standards. Their approach to safety might be driven by the desire to do only what is required to avoid citations and associated fines, and/or a desire to achieve a beneficial management system certification. Strategy-driven organizations want to achieve success in all areas of safety, not just compliance. They are focused on setting and achieving their safety goals and objectives.

Objectives are a key aspect of strategic safety. Choosing the right objectives, and their associated strategic plans, requires a detailed analysis of the organization's safety issues and gaps. Therefore, strategic safety can be defined as being analysis-driven.

Strategic safety is vision-oriented. The results of analysis helps you to envision a better future state of safety, and that vision drives you to develop strategies that best address your issues and realize your vision. Creating that vision and identifying the best strategy to realize it requires strategic thinking—thinking at a deeper and higher level.

Strategic safety is also objective-driven. An emphasis is placed on doing those things that will have the greatest impact on achieving the company's safety goals and objectives. This requires strategic planning— planning that goes beyond the typical planning used to manage safety.

Since strategic safety is focused on achieving objectives, companies must have a way to determine when their objectives are met. This requires them to develop performance measures to compare their results against. This makes strategic safety results-driven.

Another key aspect of strategic safety relates to the allocation of resources. In his book *How Successful People Think: Change Your Thinking, Change Your Life*, John C. Maxwell, a well-known American author on leadership, states: "A strategy that doesn't take into account resources is doomed to failure."[2] Companies often have limited safety staff and other resources, so they need to ensure those resources are allocated effectively.

Strategy versus Tactics

Safety and warfare may seem like strange bedfellows, but strategy and warfare are not. People have used strategy for millennia in warfare, where it originated. Even the staunchest opponents of war can learn from the strategy and tactics used in warfare. If you examine strategy from a military perspective, you can better understand it from a safety perspective. Comparing safety strategy to military strategy is not promoting or glorifying warfare, simply learning from it.

Given the term's military origin, it is helpful to distinguish strategy from tactics, which also has a military origin. *Tactics* is derived from the Greek word *taktikḗ*, which means ordering troops in combat.

In "The Origin of Strategy," Horwath summarizes these two words:

The complementary nature of strategy and tactics has defined their intertwined existence. In the military realm, tactics teach the use of armed forces in engagements, while strategy teaches the use of engagements to achieve the objectives of the war.[3]

That gives us a basic understanding of the distinction between the two terms

Strategic safety emphasizes the strategic side of safety rather than the tactical side. That is not to say that tactical safety efforts are unimportant; they *are* important. It also is not saying that companies should not do everything required of them regarding safety. What it is saying, is that companies should focus their efforts on or give priority to those tactics that will achieve their safety objectives.

Strategic safety does not ask what *can* be done to improve safety performance, but what *should* be done to achieve safety goals and objectives. The former generates a longer list of actions, but the latter produces actions that will ensure success.

Strategy & Warfare

The Levels of Warfare

To better understand strategic safety, think of it in terms of warfare. Figure 1 depicts figure I-2 in the U.S. Joint Chiefs of Staff (JCS) Joint Publication 1 (JP1), titled *Doctrine for the Armed Forces of the United States*. It is the JCS's capstone publication. That publication describes three levels of warfare.

> While the various forms and methods of warfare are ultimately expressed in concrete military action, the three levels of warfare—strategic, operational, and tactical—link tactical actions to achievement of national objectives. There are no finite limits or boundaries between these levels, but they help commanders design and synchronize operations, allocate resources, and assign tasks to the appropriate command. The strategic, operational, or tactical purpose of employment depends on the nature of the objective, mission, or task.[4]

In the JCS's figure 1-2, you can see how strategic safety adds a higher level of thought to safety. Many safety professionals and organizations approach safety at just the operational and tactical levels. It is helpful to draw parallels between strategic safety and the levels of warfare to see the benefit of adding a third, higher level to safety management.

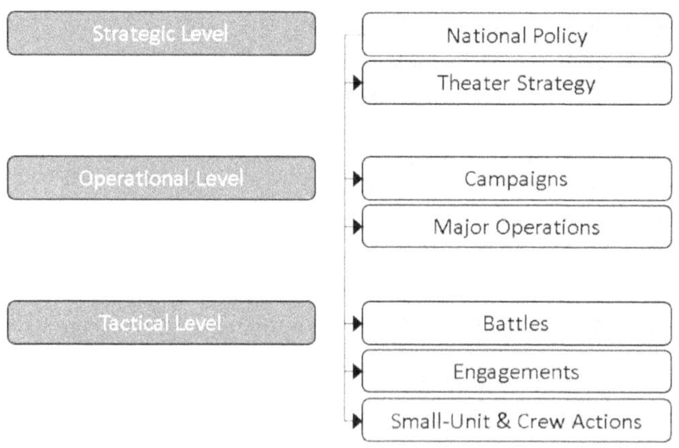

Figure 1.   Levels of Warfare (From JP1 figure I-2)

Strategic Level

The JCS doctrine includes the following in its description of the Strategic Level:

> At the strategic level, a nation often determines the national (or multinational in the case of an alliance or coalition) guidance that addresses strategic objectives in

support of strategic end states and develops and uses national resources to achieve them.[5]

Note that a nation first determines the desired end states and then develops strategies to achieve them.

Operational Level

The JCS doctrine describes the Operational Level as follows:

The operational level links strategy and tactics by establishing operational objectives needed to achieve the military end states and strategic objectives. It sequences tactical actions to achieve objectives. The focus at this level is on the planning and execution of operations using operational art...[6]

Note the intermediary level between strategy and tactics, the operational level. That level connects the other two levels and drives the lower level.

Tactical Level

The JCS doctrine states the following regarding the Tactical Level:

Tactics is the employment and ordered arrangement of forces in relation to each other. The tactical level of war is where battles and engagements are planned and executed to achieve military objectives assigned to tactical units or joint task forces (JTFs). Activities at this level focus on the ordered arrangement and maneuver of combat elements in relation to each other and enemy to achieve combat objectives.[7]

This is what Sun Tzŭ referred to as the tactics whereby he conquers—the visible aspect of war. Tactics are what we see on the nightly news or in the press about a military conflict. We do not see the strategy that led to them.

In safety, tactics are what we refer to when we talk about doing safety. They are the day-to-day activities that are driven and controlled by the company's safety plans, programs, and procedures. Tactics are how we physically achieve safety goals and objectives.

As in warfare, strategic safety incorporates strategic, operational, and tactical plans. This can be seen in addressing Serious Injuries and Fatalities (SIFs). Thought leaders and corporations are now taking a more strategic approach to addressing SIFs. While the safety community and businesses have done a good job in reducing injury rates, fatality rates have remained the same.

This prompted the safety profession to take a harder look at SIFs. Some have used the following strategy to reduce them:

Strategic Plan: Identify the causes of and precursors to SIFs, and the associated control measures. Incorporate those aspects of SIFs into programs and processes designed to prevent them.

Operational Plans:

- Research and analyze SIFs.

- Develop an SIF Prevention Program that incorporates both proactive and reactive elements.

Tactical Plans: Review available research on SIFs in order to:

- Define SIFs and establish an SIF Rate for tracking and trending.

- Analyze the company's historical SIFs for common causes and precursors.

- Develop a decision tree for determining which activities/accidents have/had the potential for an SIF.

- Develop an SIF prevention checklist for activities having SIF potential, and conduct periodic audits of those activities.

- Add a Precursor Analysis element to the Incident Investigation and Root Cause Analysis process for accidents having SIF potential.

After drawing parallels between the JCS's levels of warfare and strategic safety, it should be clear to see the distinction between strategy and tactics. With only a tactical approach, you might win some battles, but you risk losing the war. As Ho Shih warned, relying on brute strength alone will not assure victory; you need to start with strategy.

Targets & Measures

In military strategy, choosing the right targets is critical to mission success. One strategy used in modern warfare is gaining air superiority early in the conflict. Strategies involve removing the enemy's ability to defend itself and strike back by taking out its missile defense systems and anti-aircraft batteries, as well as its missile launch sites and military airfields.

You want those targets taken out in the early days of battle. Tactics used would include the use of missiles launched from naval vessels offshore, followed by the use of attack aircraft or bombers. Intelligence gathered ahead of time would provide the target locations.

There is one more aspect of military ragets, or strategic objectives—the measures of success. The measure of success is how one determines whether a strategic objective has been achieved. Here, aerial photographs, satellite imagery, and reports from secret operatives on the ground would provide military commanders with a damage assessment after the initial attacks. That information would show how well the tactics worked. A measure of success in this case would be the percentage of an enemy's air defense capabilities removed.

Why are targets and measures important to strategy? In this example, military commanders need to know when conditions are safe to deploy ground forces. Therefore, not only is it important to pick the right targets, it is important to know whether you took those targets out.

Like warfare, safety strategies must include specific targets, time frames, and measures of success. Without those elements, you do not have viable strategic objectives; all you have is wishful thinking.

The SIF example above provides a good example of choosing specific targets, SIFs, and establishing a measure for success (e.g. a 20% reduction in SIF Rate). Rather than continuing to rely on the standard safety practices for reducing general injuries, and hoping they will reduce SIFs, this strategy targets SIFs exclusively, and will help to ensure success in reducing them.

Victory versus Defeat

Strategic safety aims to be victorious. Why would anyone go to war without assurance of victory? We can ask the same about safety. Why would safety professionals want to develop plans that cannot ensure success? Unfortunately, many do, despite their well-meaning efforts.

Strategic safety goes to great lengths to identify all the key aspects of a problem and addresses them in strategic plans. When developing your strategies, have the mindset that you are going to solve the problem completely, once and for all. Seek battle only after victory is assured, by taking a comprehensive approach toward resolving the issue.

If the example of the war in Vietnam has taught us anything, it was that continuing to escalate a war without a clear and viable strategy for victory, is a recipe for failure. Even though tactics like the massive bombing campaign known as Operation Rolling Thunder, and the ground tactic referred to as search and destroy, were successful in their own right, they ultimately did not lead to victory. In the end, the war's primary objective, to prevent the spread of communism in Southeast Asia, was not achieved.

Similarly, in safety, we see organizations adding tactic after tactic to resolve safety performance issues, only to effect minimal change. If you are going to do something, do it right the first time. Don't just try something and hope it works.

Our data-driven, targeted approach to reducing hand injuries in the new joint venture, referenced in the introduction, was a strategy with victory in mind. The ultimate objective was to eliminate specific types of hand injuries. We had no hand-related injuries involving the use of hammers and striking implements, or while cutting using knifes, save one, for years following the implementation of that strategy.

Course Corrections

Stating that strategy wins the war does not diminish the role that operational and tactical plans play in its execution and success. We all know plans change. You may have heard: "No plan survives the first contact with the enemy." This is a paraphrase of what Prussian Field Marshal Helmuth von Moltke the Elder wrote about war strategy and operational plans.

> The tactical result of an engagement forms the base for new strategic decisions because victory or defeat in a battle changes the situation to such a degree that no human acumen is able to see beyond the first battle. In this sense one should understand Napoleon's saying: "I have never had a plan of operations." **Therefore, no plan of operations extends with any certainty beyond the first contact with the main hostile force.** Only the layman thinks that he can see in the course of the campaign the consequent execution of an original idea with all details thought out in advance and adhered to until the very end.[8] (Emphasis Added)

The same holds true with the operational and tactical plans developed as part of a safety strategy. Circumstances can change throughout execution. If a course correction is needed, you need to make it.

> Sun Tzŭ likened tactics to water, and how the changing terrain dictates its flow.

> Water shapes its course according to the nature of the ground over which it flows; the soldier works out his victory in relation to the foe in whom he is facing. Therefore, just as water retains no constant shape, so in warfare there are no constant conditions. He who can modify his tactics in relation to his opponent and thereby succeed in winning, may be called a heaven-born captain.[9]

If you have been managing safety for any length of time, you know how the safety landscape at work changes on a regular basis. To keep your strategy steadily moving forward, you have to be fluid while executing it to account for the changing landscape.

Strategic safety is a deeper and higher level of thought pertaining to safety. Deeper as it relates to strategic thinking and analysis, and higher as it relates to strategic planning. Strategic safety does not emphasize a certain aspect of safety; it emphasizes achieving safety goals and objectives. Strategic safety answers the fundamental questions: "Where are we, where do we want to go, and how do we get there?"

Strategic safety can be described as an approach to safety that is:

- Vision-oriented, using strategic thinking.

- Analysis-driven, using strategic analysis.

- Objective-driven, using strategic planning.

- Results-driven, using performance measures.

Describing it further, strategic safety:

- Addresses targets that will have the most impact on achieving safety goals and objectives.

- Uses measures to determine the level of success in achieving safety goals and objectives.

- Allocates resources in the most effective way to achieve safety goals and objectives.

- Seeks to be victorious, resolving situations or achieving objectives to their fullest.

- Is dynamic, monitoring progress during execution and making course corrections as needed.

Armed with a better understanding of what strategic safety is, the next order of business is to determine where to focus your efforts. Which area(s) of safety should you develop your strategic objectives in? The next chapter will help you focus your efforts where they will have the greatest impact on overall safety.

Endnotes

1. *LibQuotes*, s.v. "Voltaire Quote," accessed January 3, 2022, https://libquotes.com/voltaire/quote/lby-6f6v.

2. John C. Maxwell, *How Successful People Think: Change Your Thinking*, Change Your Life (New York, Center Street, 2009), 56.

3. Rich Horwath, "The Origin of Strategy," *Strategic Thinking Institute*, accessed October 13, 2023, https://www.strategyskills.com/the-origin-of-strategy/.

4. U.S. Department of Defense, Joint Chiefs of Staff, Joint Publication 1, Doctrine for the Armed Forces of the United States (Washington, DC: Department of Defense, July 12, 2017), I-7, https://irp.fas.org/doddir/dod/jp1.pdf

5. Ibid.

6. Ibid., I-8.

7. Ibid.

8. Helmuth von Moltke, "Plan of Operations" (1871-81)," in Moltke on the Art of War: Selected Writings, ed. Daniel J. Hughes (New York: Ballantine Books, 1993), 92.

9. Lionel Giles, *Sun Tzŭ on the Art of War.* [1910; Project Gutenberg, 2004], Chapter VI, Paragraphs 31-33., https://www.gutenberg.org/files/132/132-h/132-h.htm.

Chapter Three:
Strategic Areas of Safety

Focus areas are the foundation stones of your strategy. — Tom Wright

In modern warfare, a nation engages the enemy in four arenas: at sea, on land, in the air, and, more recently, in space. Objectives would be developed for those arenas as part of a military strategy, along with corresponding strategic, operational, and tactical plans. In the safety profession, we also operate in different arenas, focus areas within which we can develop strategies. I refer to these focus areas as the Strategic Areas of Safety.

Focus Areas

Tom Wright is the founder and CEO of Cascade, an Australian-based software as a service (SaaS) company hosting a cloud-based strategy execution platform. In an online article titled "Strategic Focus Areas: How to create them + Examples," Wright states: "Focus areas are the foundation stones of your strategy. They expand on your Vision Statement and start to create some structure around how to actually get your organization to achieve its goals."[1]

Foundation stones in masonry construction of old were used to support the building, and you know the fate of a house built on a weak foundation—it will not stand.

Cornerstones were the first foundation stones laid, and provided a reference point from which all remaining foundation stones were aligned. Think of the Strategic Areas of Safety as the cornerstones of strategic safety. They provide the reference points for your overall strategy. Like the four arenas in warfare, each strategic area would have strategic objectives and associated plans aligned under it.

A common question regarding focus areas is, "How many should we have?" In its e-book titled *How to Write a Strategic Plan: Cascade Model,* Cascade states: "We usually suggest creating between 3 to 5 Focus Areas. Any fewer and they will probably be too vague. Any more, and well…I for one certainly can't focus on more than 5 things at once!"[2]

Blue Ocean Strategies: How to Create Uncontested Market Space and Make the Competition Irrelevant is a popular book on strategy by authors W. Chan Kim and Renée Mauborgne. It presents what the authors call "The Four Actions Framework."[3] The use of four Strategic Areas of Safety fits within Cascade's guidelines and aligns with this and other popular strategy development methods.

The Balanced Scorecard method for approaching strategy is what many businesses use today. It is one of the most popular strategy performance management tools. It was introduced in the mid-1990s to help businesses develop a mixture of financial and non-financial measures from which to track and monitor business performance. The original Balanced Scorecard method viewed a business from four different perspectives: *financial, internal business processes, customer,* and *learning and growth.* These four perspectives are the cornerstones of the method and represent focus areas within which strategies and performance measures are developed.

Like the four arenas of warfare, "The Four Actions Framework," and the four Balanced Scorecard perspectives, there are four focus areas that are well-suited for the field of safety, or what I call the Strategic Areas of Safety. As shown in figure 2, these areas are *performance, compliance, competence,* and *culture.*

Figure 2.   Strategic Areas of Safety

Safety is like a two-sided coin with *performance* as heads and *compliance* as tails. You must use your head ("Think Safety") to prevent or avoid incidents. You must also cover your tail by complying with regulations. These are the primary strategic areas.

Consider these four Strategic Areas of Safety as cornerstones of strategic safety. It is within these areas that safety professionals can focus their strategic thinking, analysis, and planning efforts to achieve safety excellence. Below, we will examine each area and introduce assessments used to analyze them. The assessments will be fully explained and demonstrated in Part III.

Safety Performance

Safety performance as a strategic area encompasses what many consider the essence of safety. What is the essence of safety? Fred A. Manuele is the president of Hazards, Ltd., a well-known author on safety management, and one of the safety profession's award-winning thought leaders. In a *Professional Safety* article titled "Preventing Serious Injuries & Fatalities," Manuele stated: "The entirety of purpose of those responsible for safety, regardless of their titles, is to identify, evaluate, and eliminate or control hazards so that the risks deriving from those hazards are acceptable."[4]

Hazards are something that can cause danger, harm, or loss. Risk is the likelihood of injury, damage, loss, or any other detrimental outcome resulting from exposure to hazards. The essence of occupational safety, therefore, can be summarized as the elimination or mitigation of risk.

This strategic area focuses on activities that eliminate or mitigate risks It answers the questions: "How safe are we, how safe do we want to be, and how can we be safer?"

Safety performance analysis determines how safe the organization is, and what needs to be done to make it safer. A foundational assessment in this strategic area is a Risk Profile or Risk Registry. It identifies all the potential environmental, health, and safety risks associated with the business's facilities and operations, and determines the control measures needed for each. It is a prospective assessment.

A second assessment is an Attribute Analysis. It looks at the historical incidents experienced by the organization to identify the trends and problem areas that need to be addressed to improve safety performance. It is a retrospective assessment.

Safety Compliance

The second strategic area is compliance. Companies have various EH&S compliance requirements they must meet, including mandatory and voluntary compliance with external and internal requirements.

The biggest related concern for many safety professionals is compliance with regulations. However, there are other areas, such as compliance with international standards adopted, as well as compliance with internal EH&S requirements. Companies might also need to comply with the EH&S requirements of their customers, depending on their industry.

One responsibility of safety professionals, therefore, is to manage and promote compliance with all of those requirements. A failure to comply can cause citations and fines, loss of certifications, or loss of customers, all of which represent a potential loss to the company, either a reputational, financial, or operational loss. The compliance strategic area focuses on those aspects of safety that involve meeting compliance requirements and the measurement of their success. It answers the questions: "How compliant are we, how compliant do we want to be, and how can we be more compliant?"

Safety compliance analysis determines how compliant the organization is. One foundational assessment in this strategic area is a Regulatory Profile or Regulatory

Register. It identifies all the organization's EH&S compliance requirements. It is a prospective assessment.

Another assessment, a Compliance Task Profile, picks up where the Regulatory Profile leaves off to identify the specific compliance tasks required to be performed. It is also a prospective assessment.

A third assessment is another Attribute Analysis. This analysis looks at the historical lack of compliance presented in either external regulatory inspections or internal audits and inspections. Again, it identifies the trends and problem areas that need to be addressed to improve safety compliance. It is a retrospective assessment.

Safety Competence

The third strategic area is competence. It is one of the underlying factors driving the safety performance and compliance of the organization. Every employee at each level of the organization plays a role in it. Safety is not just the responsibility of the safety department personnel. Safety is everyone's responsibility, regardless of their role in the company. Do employees have the knowledge, skills, and abilities (KSAs) to perform safely in their individual roles, or to lead or manage safety? If not, how can you develop their competencies?

Safety competence is not something that is taught in a training session, it is developed. The competence strategic area focuses on those aspects of safety that are involved with developing a higher level of safety competency within the organization. It answers the questions, "How competent are our people, how competent do we want them to be, and how can we improve their competency?"

Safety competence analysis determines how competent the organization is in fulfilling their safety responsibilities. One assessment in this strategic area is a Safety Training Profile. It examines the safety training requirements for all applicable positions within the organization, based on external and internal requirements. It is a prospective assessment.

Another assessment is a Competency Assessment, which evaluates the competency of safety professionals and other key positions in the organization. These assessments help to determine the current level of safety qualification and proficiency, and identify areas where improvements are needed to raise the level of safety competency.

Safety Culture

The fourth strategic area is culture. It is the most important underlying factor driving the organization's safety performance and compliance. Stephen I. Simon, Ph.D., president of Culture Change Consultants, Inc., regarded as the father of safety culture, defines culture as follows:

> Every organization has a safety culture. Culture matters because it strongly influences how we think, what we feel, and how we behave. A formal definition of culture is the set of basic assumptions, perceptions, values, and beliefs a group makes about their reality in their particular universe. An informal definition of culture is 'what happens when no one is watching.'[5]

Every organization has a safety culture whether or not they manage it. Organizations that have a positive, proactive, participative safety culture have cultivated it to be that way. The culture strategic area focuses on those aspects of safety that are involved with cultivating and maintaining the safety culture. It answers the questions, "How good is our culture, how good do we want our culture to be, and how can we improve it?"

Safety culture analysis evaluates the maturity and current state of the organization's safety culture. One assessment is a Maturity Assessment, which analyzes the maturity level of your overall safety program and/or safety culture as it progresses through various stages of maturity.

Another assessment is the Safety Perception Survey. This assessment solicits feedback about the current state of the organization's culture, its climate, from managers, supervisors, and front-line employees.

Safety strategies have focus areas that each demand asking these questions. "Where are we, where do we want to go, and how do we get there?" Finding these answers requires strategic thinking and analysis, two primary aspects of strategic safety. These answers provide the basis for safety strategies, which are developed through strategic planning.

Part I introduced the concepts of strategy and strategic safety, with its four strategic areas. Parts II through IV will examine the following three aspects of strategic safety in greater detail: *strategic thinking*, *strategic analysis*, and *strategic planning*, respectively.

Endnotes

1. Tom Wright, "Strategic Focus Areas: How to create them + Examples," *Cascade*, accessed December 2, 2023, https://www.cascade.app/blog/strategic-focus-areas.

2. "How to Write a Strategic Plan: Cascade Model," *Cascade*, 9, accessed February 7, 2022, https://www.cascade.app/strategic-planning-models-ebook.

3. W. Chan Kim and Renée Mauborgne, *Blue Ocean Strategies*, (Boston: Harvard Business School Press,20052), 29.

4. Fred A, Manuele, "Preventing Serious Injuries & Fatalities," *Professional Safety*, May 2013, 55.

5. Stephen I. Simon, Ph.D., "Transforming Your Safety Culture: How You Can Start to Make it Happen," *Culture Change Consultants*, 1, accessed February 8, 2022, https://www.culturechange.com/wp-content/uploads/2018/10/Transforming-Your-Safety-Culture.pdf.

Part II
Strategic Thinking

Chapter Four:
Generating Insights

Analysis is the critical starting point of strategic thinking. — Kenichi Ohmae

Now that you have reached Part II of this book, it is time to put on your thinking cap—your strategic thinking cap. If you are a fan of wearing ball caps, chances are you have a selection to choose from. The cap you choose to wear is probably dependent on the context in which you are wearing it. The same is true for thinking caps, the imaginary caps you wear to do different types of thinking, such as analytical, critical, creative, or strategic, to name a few.

Henry Mintzberg, a Canadian academic, is one of the most cited authors on business strategy. In his book *Strategy Safari: A Guided Tour Through the Wilds of Strategic Management*, he stated that "Research tells us that…strategy-making is an immensely complex process involving the most sophisticated, subtle, and at times subconscious of human cognitive and social processes."[1]

Do not let Mintzberg scare you. His description may be correct, but it does not mean that strategy is beyond your grasp. These sophisticated, subtle, and at times subconscious cognitive processes can be studied and practiced to develop your strategic thinking skills. This chapter will demystify the concept of strategic thinking by providing you with a mental construct from which to better understand it.

What Is Strategic Thinking?

This and the following chapter examine the three aspects of the strategic thinking process: *information gathering, information analysis* and *strategy formulation.*

What exactly is strategic thinking? In a white paper titled "Are You Insightful or Insight-less," Rich Horwath, CEO of the Strategic Thinking Institute, states, "Strategic thinking is defined as 'the ability to generate insights on a continual basis to achieve competitive advantage.'"[2] Since safety professionals develop functional or departmental strategies, as opposed to business strategies, we can redefine it as "the ability to generate insights on a continual basis to achieve safety excellence."

The Strategic Thinking Process

In a white paper titled "The Skills of Strategic Thinking," Horwath states: "At the strategic level, content expertise alone is not sufficient. Strategic thinking is process-based, rather than content-based. In every situation where strategy is involved, there are two dynamics at work—process and content."[3] Safety professionals are content experts, but they also need to know the strategic thinking process when it comes to strategy.

In safety, we have another process similar to the strategic thinking process that can help us better understand it. That process is Incident Investigation and Root Cause Analysis (i.e. The RCA process). The RCA process has three core phases: *fact-finding, causal analysis,* and *preventive actions.* The three phases of the strategic thinking process are similar to them. Figure 3 uses the shape of an hourglass to represent that process.

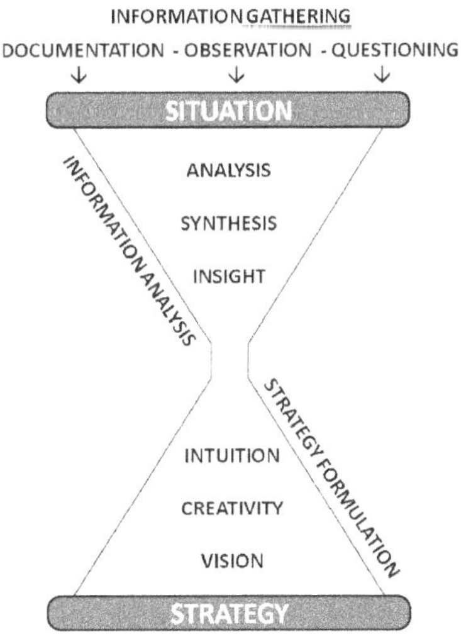

Figure 3.   Strategic Thinking Hourglass

A mining and manufacturing analogy helps describe this process. Ore is gathered from the mine. It is then placed in the top chamber for refining through analysis and synthesis, generating insights about the situation. These insights—now refined materials—transfer to the bottom chamber where a vision is forged through intuition and creativity before being framed in a way that will drive an executable strategy.

Like the RCA process, the strategic thinking process can range from simple to complex. Even though the same process is used for both, an RCA for a fatality is more complex than for a near miss. The same holds true for strategic thinking; its level of complexity depends on the situation. In this chapter, and the next, we will use an example more focused in nature, reducing injury rates, namely hand injury rates.

The strategic thinking process is just that, a process. It is not a strict step-by-step procedure. As you progress through it, like during the RCA process, you gain more insight. This often requires you to gather more information, things you did not initially consider. Strategic thinking is a process, and processes are more dynamic than static.

Phase 1 - Information Gathering

The fact-finding phase of the RCA process involves reviewing documentation, inspecting the incident scene, and interviewing witnesses. Gathering information for the strategic thinking process includes three similar components: *documentation*, *observation*, and *questioning*.

Documentation

When gathering information for the strategic thinking process, it is best to start with all documentation associated with the situation. Documentation will help inform you of what observations and questions are needed.

The specific documentation needed depends on the situation. It might include written policies, programs, and procedures, information regarding incidents and injuries, or training and other records. Gather all documentation regardless of its relevance. The analysis phase will determine if it is relevant and meaningful, or could help generate further insight into the situation.

It is said that the most important aspects of real estate are location, location, location. Likewise, three of the most important aspects of safety are documentation, documentation, documentation. If your company were experiencing an increased trend in injuries, you would start by gathering reports from the injury reporting database and the individual injury reports. With strategic thinking, documentation like that provides valuable raw material for analyzing the situation.

Observations

As safety professionals, we use observational skills to identify unsafe acts and unsafe conditions. We use these skills when we perform a safety observation, conduct a safety inspection, or do a Job Safety Analysis (JSA), among others. We also use observational skills when performing incident scene inspections during an RCA.

Those same skills, albeit applied with a strategic mindset, are a valuable part of the strategic thinking process. While the observational skills used in other safety processes are like those used in strategic thinking, the mindset must be different. Rather than looking for unsafe acts or conditions, the observer is trying to gain insight. The strategic thinking process is looking for different ways of doing something that will help to address the situation, and that requires gaining a deeper level of insight into it.

The important part is that strategic thinkers need to observe things for themselves, whenever possible. Firsthand knowledge is important during the remaining phases.

Questions

Like observational skills, safety professionals also use questioning skills. We use these skills during witness interviews as part of an RCA, for example. Many of the same principles and skills used during witness interviews apply to questioning for strategic thinking, albeit with a different focus.

In the RCA process, you ask questions to get the facts related to the incident. It is like the old radio and television series, *Dragnet*, where Sgt. Joe Friday said: "All we want are the facts, ma'am." In the strategic thinking process, you ask questions to get broader information about the situation, which also includes opinions and perceptions, among others.

Strategic thinking also requires a higher level of questioning. This includes questions about the overall context, and questions that make you think about your own assumptions.

Below are some examples of higher-level, more strategic questions.

- What is the critical issue involved?
- What is the ultimate objective we are trying to achieve?
- Is the current approach being taken the most effective one?
- What assumptions are we making, and are they valid?
- Is this a resource, systems, or human performance issue?
- Do our safety management systems address the issue?

In the information-gathering phase you need to gather documentation, make first-hand observations, and question the people involved. Your aim is to gather sufficient information for the next phase. The more information you have, the more complete your analysis is. The more analysis results you have, the more insight you will gain for strategy formulation.

Phase 2 - Information Analysis

Kenichi Ohmae (Mr. Strategy), in his seminal book *The Mind of The Strategist: The Art of Japanese Business*, writes:

> Analysis is the critical starting point of strategic thinking. Faced with problems, trends, events, or situations that appear to constitute a harmonious whole or come packaged as a whole by common sense of the day, the strategic thinker dissects them into their constituent parts. Then, having discovered the significance of these constituents, he reassembles them in a way calculated to maximize his advantage.[4]

The information analysis phase is the more active phase of strategic thinking, involving: *analysis*, *synthesis*, and *insight*. First you analyze the information and break it down into its constituent parts, then you synthesize those parts to create a new way of thinking about them to gain insights. These insights will then help you formulate your solution or strategy.

Analysis

Ohmae states:

> The first stage in strategic thinking is to pinpoint the critical issue in the situation. Everyone facing a problem naturally tries in his or her own way to penetrate to the key issue. Some may think that one way is as good as another and that whether their efforts hit the mark is largely a matter of luck. I believe it is not a question of luck at all but of attitude and method.[5]

A method that Ohmae recommends for getting to the critical issue is what he calls an Issue Diagram. This is simply a tree diagram, similar to a causal tree diagram, that uses questions to examine the various aspects of the broader issue. The key is asking questions with a yes or no answer to guide you to the underlying or critical issue. It helps you to clearly define a problem.

Figure 4 provides a simple example of the start of an Issue Diagram for developing an injury reduction strategy.

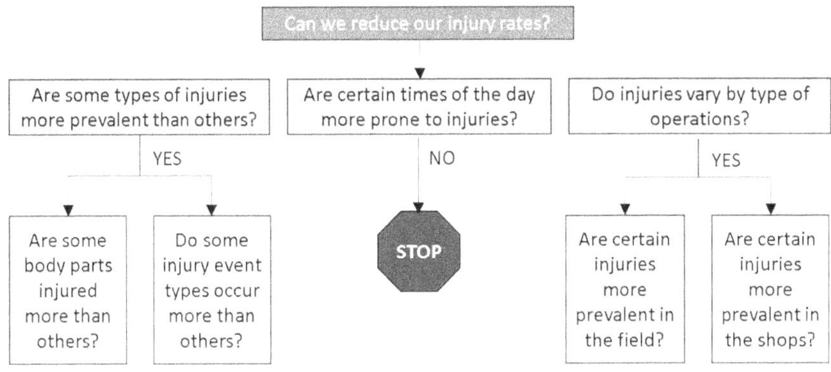

Figure 4.   Issue Diagram

The Issue Diagram serves a similar function as the Five Why Analysis, the basic principle behind many RCA methodologies. As you know, the Five Why Analysis involves asking why an incident occurred through at least five levels to uncover the root cause. Here, you also ask more questions to dive deeper.

An Attribute Analysis of injuries is needed during the information-gathering phase to complete the Issue Diagram above. An Attribute Analysis, described in more detail in Chapter Six, comprises analyzing various attributes, such as body parts or injury type, to help narrow down the problem or identify the critical issue(s).

Table 1 provides a simple example of an Attribute Analysis used during the information analysis phase to answer some of the injury-related questions raised in the Issue Diagram. It allows the analyst to count the attributes in each category to better understand the overall injury information. You can perform this analysis before or during the completion of an Issue Diagram, but without this data you would be hard pressed to actually complete the diagram.

Table 1.   Injury Attribute Table

1	Shop	Hand	Cut	Struck By	Knife
2	Field	Hand	Contusion	Struck By	Hammer
3	Office	Foot	Sprain	Slip & Fall	N/A
4	Field	Hand	Fracture	Struck By	Hammer
5	Shop	Hand	Cut	Struck By	Knife
6	Shop	Back	Sprain	Lifting	N/A
7	Field	Hand	Fracture	Struck By	Hammer

How does this data help? In the table above, hand injuries account for five out of seven (71%) of the injuries. Under the injury event attribute, struck by accounted for the same number of injuries. The tools personnel were struck by include hammers and knives. Neither field nor shop injuries seem to be more prevalent. Injury types and causal themes vary. Day of the week or time of day do not appear to be a factor in the injuries.

This Attribute Analysis helped to identify the critical issue. Employees are striking and injuring their hands while using hammers in the field and knives in the shops. Combining that with other results from your analysis, will help you gain better insight from which to formulate your strategy.

The benefit of using formal analysis techniques, such as those described above, is that they help you go beyond a surface-level understanding of the issue. Strategic thinkers dive deeper to find what is at the heart of the problem. As the old saying goes, "the devil is in the details."

In a Harvard Business Review article titled "The Fall and Rise of Strategic Planning," Mintzberg wrote: "Real strategists get their hands dirty digging for ideas, and real strategies are built from the nuggets they uncover." [6]

In his paper titled "Deep Dive: The Three Disciplines of Strategic Thinking," Horwath uses the analogy of underwater diving to describe four types of strategic thinkers. He states that the best strategic thinkers are those who can dive deep. In setting up that analogy, Horwath states: "Herein lies the pearl of great opportunity: the deeper you can dive into the business and resurface with strategic insights, the more valuable you'll become to your organization."[7]

As an ex-submariner, I remember hearing the dive command and alarm coming over the 1MC from the Chief of the Boat: "Dive, dive, (ah-OOG-ah, ah-OOG-ah), dive, dive." At least once during each patrol we dove really deep, going to the vessel's test depth. We did that to test the condition of systems designed to keep seawater out of the boat, while under higher sea pressure, gaining valuable insight into the boat's condition. It was a little eerie, but so long as the number of surfaces equaled the number of dives, we were okay.

Diving deeper during your strategic thinking information analysis, like going to test depth on a submarine, helps you to see things not evident while on the surface. While you are diving, be sure to include looking at your safety management systems. Put them to the pressure test to see if any improvements are needed to avoid future issues.

Synthesis

Synthesis combines two or more ideas, concepts, or elements to form a new idea, theory, or plan. It relates to the term synergy, which is a type of synthesis. Synergy produces a combined effect greater than the sum of the separate effects from the individual ideas.

As safety professionals, we understand the synergistic effect in terms of chemistry. Consider the childhood baking soda and vinegar volcano science experiment. Both substances are chemicals with separate and milder effects. When you mix the two, however, they produce a dramatic bubbling effect that simulates a volcanic eruption.

Synthesis, then, is combining ideas to form something new or novel. One example of synthesis occurred while writing this chapter. While conducting my research, the elements of strategic thinking started looking similar to the elements of the RCA process. That prompted the development of the three phases used here to describe the strategic thinking process. Two concepts were combined (synthesized) to form a novel way of explaining strategic thinking that safety professionals could relate to.

Synthesis in strategic thinking is not always that intuitive or simple. It often takes a deeper examination of the information. There are techniques and tools that aid in the process, however. One technique is Mind-Mapping.

Mind Maps were first introduced by Tony Buzan, known as the father of mind mapping, during the 1970s. It is a radial diagram based in part on concept maps, and is used as a creative thinking tool. Starting with the situation in the center, this technique places information by topic in concentric tree diagrams as shown in figure 5.

With the problem at the center, the map then branches out using various aspects of the problem. Once finished, the map enables a comparison of information from the various branches to identify connections or similarities. Keywords or phrases are used to help make those connections. These keywords include "causes," "requires," "contributes to," or "is like."

While there are software applications available for mind mapping that provide advanced graphics, it can also be performed on a whiteboard or by using Post-it notes on a wall, such as during a brainstorming session.

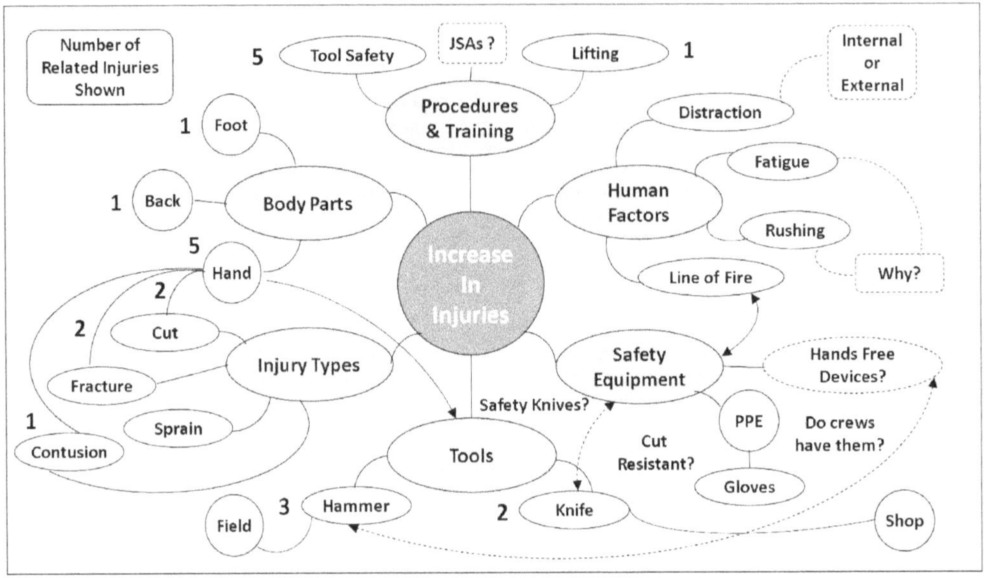

Figure 5.   Injury Mind Map

In the Injury Mind Map above, you will see connections highlighted using dotted lines. They are basic examples of synthesizing data, such as between the Tools and Safety Equipment branches. They present questions that, when answered, could provide useful insights for developing a strategy to reduce injuries.

One of the benefits then, of Mind Maps, is that they help you to avoid siloed thinking when it comes to where a solution might be found. They help you to see the problem from different perspectives, each of which could contribute to a solution through synthesis.

Insight

Insight is defined as an instance of having a clearer understanding of something and discerning its true nature. It is also the ability to gain that understanding or discernment, especially through intuition. In "The Strategic Thinking Manifesto," Horwath states: "The common core of both strategy and innovation is insight. An insight results from the combination of two or more pieces of information or data in a unique way that leads to new value for customers."[8]

As safety professionals, we often find ourselves in situations where we need to solve problems or drive improvements. In both cases, we need the best possible insight to achieve our objectives. The strategic thinking process helps you generate insights.

Not all insights may be useful in helping you to achieve your objectives. Two key factors in determining the usefulness of an insight are *relevancy* and *meaningfulness*. You should ask yourself whether the insight is both relevant and meaningful.

1. *Relevant:* An insight is relevant if it pertains to your objective, either the current state, the desired end state, or the plan to get you there. The fundamental strategy questions are, "What is the current state, what is the desired state, and how can we best get from here to there?" Does the insight help you answer either of those questions in any way?

2. *Meaningful:* An insight is meaningful if it has importance or value related to your objective. An insight might pertain to your objective, but does it have enough energy or force to help drive the needle in achieving your objective?

The example Mind Map included three abbreviated questions. The answers to the questions came from further information gathering.

1. *Does the company use safety knives?* No, the company does not provide safety knives. Those are knives that either provide a guard over the blade or retract the blade once it leaves the material being cut.

2. *Does the company provide cut-resistant gloves?* Yes, the company provides cut-resistant gloves in the shops. However, there is no mechanism in place to state which tasks those gloves are required to be used for, and the cut-resistance rating required.

3. *Do crews have hands-free devices?* No, the field crews do not have hands-free devices. They use ropes sometimes to hold the slugging wrench in place, while a sledgehammer is being swung at it, so they recognize the need for such devices.

These are some of the critical issues related to the increase in injuries. Although insights in and of themselves, they could also provide additional insight to help generate an injury reduction strategy.

Recall that insight is the ability to gain a clearer understanding of something, discerning its true nature. Recall also that strategic thinkers dive deeper to find valuable insight. With that in mind, let us dive deeper to gain further insight beyond merely answering those questions. In doing so, it helps to apply the five why principle to gain a deeper understanding.

First, the company does not appear to be providing safety equipment that could prevent hammer and knife-related injuries. Why is that? Perhaps the Job Hazard Analysis (JHA) process was not applied to those tasks, or the person(s) who performed the JHAs were not aware of those safety devices. This might suggest a lack of competencies in the persons performing JHAs. Maybe only front-line workers perform JHAs, rather than involving safety professionals.

Another insight is the fact that the RCA process performed on similar past injuries failed to consider the use of those safety devices. Why is that? Similar to JHAs, it could be that the person(s) who performed the RCAs were not aware of those safety devices. If they were familiar with them, did they create a preventive action to obtain those devices? If they did, why were the actions not completed?

Whatever the reason, the RCA process failed to prevent future injuries of that type. When performed right, the RCA process is a valuable tool for reducing injuries. When it doesn't, you have to evaluate whether the tool is being used correctly, or if the RCA methodologies used are effective enough.

This brings to mind the discussion about victory versus defeat in Chapter Two. If you truly want to eliminate safety problems, like injuries, continue to dive deeper into the problem to get all of the insight you can about it. When an insight raises more questions, that's your cue to keep diving deeper.

Rather than focusing on surface level solutions, strategic thinking should drive you toward more comprehensive solutions. The same is true of the RCA process. When I perform an RCA, I look at both contributing and causal factors. Addressing both can help prevent future incidents. As for root causes, I am not a member of the "there is only one root cause" club. While there may be a predominant root cause, every causal chain leads to a root cause, and multiple chains are often involved. Again, addressing each of them helps to prevent future incidents.

The hand injury example showed how strategic thinking can generate insights for the strategy formulation phase. The more insight you gain, the better you strategy will be. Some of the insights generated will provide material for your strategic, operational and tactical plans, while others will help you develop higher level strategic objectives.

We compared the information-gathering phase to mining natural resources, and the information analysis phase to a refining process. The better the refining process is, the better quality the manufactured product will be. This is true of insights generated through strategic thinking, as well. Strategic thinking is a process very similar to the RCA process. It begins with gathering information about a situation, and then analyzing that information to gain more insight into it.

In this chapter, I covered these first two phases of the strategic thinking process. In the next chapter, I will cover the third phase of the process, strategy formulation. In doing so, I will continue using the same hand injury example. By learning and applying this process, you increase the likelihood of developing strategies that will achieve your goals or objectives.

Endnotes

1. Henry Mintzberg, *Strategy Safari: A Guided Tour Through the Wilds of Strategic Management* (New York, The Free Press, 1998), 73.

2. Rich Horwath, "Are You Insightful or Insight-less?" *Strategic Thinking Institute*, accessed February 1, 2022, https://www.strategyskills.com/free-resources/white-papers/are-you-insightful-or-insight-less/

3. Rich Horwath, "The Skills of Strategic Thinking," *Strategic Thinking Institute*, accessed

4. Kenichi Ohmae, *The Mind of the Strategist: The Art of Japanese Business* (New York: McGraw-Hill,1982), 12.

5. Ibid., 15.

6. Henry Mintzberg, "The Fall and Rise of Strategic Planning," *Harvard Business Review*, accessed January 22, 2022, https://hbr.org/1994/01/the-fall-and-rise-of-strategic-planning.

7. Rich Horwath, "Deep Dive: The Three Disciplines of Strategic Thinking," *Strategic Thinking Institute*, accessed January 20, 2022, https://www.strategyskills.com/wp-content/uploads/Deep-Dive.pdf.

8. Rich Horwath, "The Strategic Thinking Manifesto," *Strategic Thinking Institute*, accessed January 20, 2022, https://www.strategyskills.com/wp-content/uploads/2012/09/The-Strategic-Thinking-Manifesto.pdf.

Chapter Five:
Creating a Vision

It's an open secret that good ideas come to you as flashes of insight, often when you don't expect them. — William Duggan

After gathering and analyzing the relevant information to generate insights, what do you do with those insights? Well, keep your strategic thinking cap on, it's time to create your vision —your strategic vision. Without your strategic thinking cap, you might be tempted to just write a few action plans. With it, however, you will generate something more effective and lasting, a strategic vision.

In this phase, we will examine three more elements of strategic thinking: *intuition, creativity,* and *vision.*

Strategic thinking involves intuition and creativity, both of which fit what Mintzberg described as sophisticated, subtle, and at times subconscious human cognitive processes. If you do not view yourself as being intuitive or creative, fear not. We will examine these elements of strategic thinking, and learn some ways to help you use them to think more strategically.

Phase 3 - Strategy Formulation

Here we begin operating in the bottom half of the strategic thinking hourglass, the forge in our mining analogy. As more and more insights drop down from the top half, we begin to start forming a strategic vision.

Intuition

Intuition is the ability to understand something without the need for logic, analysis, or conscious reasoning. It is a natural ability to know something seemingly having no proof or evidence.

Psychologists and neuroscientists both agree that our minds have two cognitive modes, analytical and intuitive. The analytical mode is conscious, rational, and deliberate. The intuitive mode is subconscious, experiential, and automatic. We have already discussed the analysis phase of strategic thinking, which would use the analytical mode. Here we will examine the intuitive mode and how it contributes to strategic thinking.

William Duggan is an academic, researcher, and author. He is a senior lecturer in business at the Columbia Business School and has twenty years of experience as a strategy adviser and consultant. Duggan begins his book *Strategic Intuition: The Creative Spark in Human Achievement*, by introducing a new discipline he calls *strategic intuition*:

> It's an open secret that good ideas come to you as flashes of insight, often when you don't expect them. It may have happened to you—in the shower, or stepping onto a train, or stuck in traffic, falling asleep, swimming, or brushing your teeth in the morning. Then it hits you. It all comes together in your mind. You connect the dots. It can be one big "Aha!" or a series of smaller ones that together show you the way ahead. The fog clears and you see what to do. It seems so obvious. A moment before you had no idea. Now you do.[1]

Duggan says those flashes of insight, those "Aha!" moments, represent a distinct form of intuition—*strategic intuition*—which falls under the field of strategy. Duggan describes three types of intuition: *ordinary*, *expert*, and *strategic*.

1. *Ordinary Intuition:* A vague hunch or gut instinct; feeling, not thinking.

2. *Expert Intuition:* Jumping to a conclusion; thinking, not feeling; always fast and works when something is familiar.

3. *Strategic Intuition:* A flash of insight or Aha! moment; thinking, not feeling; always slow and works when something is new or unfamiliar.[2]

Intuition involves recalling memories stored in the mind. Those memories can be from personal experiences, which are stronger and more easily recalled. They can also be from the experience or knowledge of others. Those memories are weaker and more difficult to recall. The recall in this case involves associations between multiple memories stored in different parts of the brain, hence the slower process. Expert intuition stems from personal experience, whereas strategic intuition stems from both personal experience and the experience of others.

Using the increase in injuries examples from the previous chapter, we can better understand the difference between expert and strategic intuition.

1. *Expert Intuition:* Suppose you introduced safety knives at a previous job. Therefore, you have first-hand knowledge of how they can prevent lacerations during cutting tasks. After learning that the company did not use safety knives, implementing their use readily came to mind. This is expert intuition.

2. *Strategic Intuition:* Suppose you never used hands-free devices for hammering tasks. You saw employees putting their hands in the line of fire. A solution for it did not immediately come to mind based on your personal experience alone. At some point, a solution comes to mind when things you had seen, read, and/or heard from others gets added to the mix. This is strategic intuition.

Imagine this scenario occurring after completing your first draft of the Mind Map. You are at home hanging some pictures. When you start hammering an anchor in the wall, a flash of insight hits you. You recall a seminar you attended a few years back where a group of safety professionals from the oil and gas industry were talking about "Fingersavers." Fingersavers are hands-free devices that were designed for an oil refinery in the United Kingdom to help reduce hand injuries. Rather than holding a slugging wrench with your hand, putting it in the line of fire of the sledge hammer, the Fingersaver was used to hold it.

Strategic intuition, in this case, combined personal knowledge about line of fire hazards, and the knowledge of others that you gained a few years ago during a seminar. Because the memory of that seminar discussion was deeply buried, it took a while to recall it. When you placed your own fingers in the line of fire, it helped trigger that memory. Thanks to the insight you had recently generated, you began to formulate a strategy.

In his book, Duggan states that expert intuition can be the enemy of strategic intuition. This occurs when someone jumps to a conclusion and makes a snap judgment based on personal experience alone. Strategic intuition is more impactful because it incorporates the knowledge of others.

Duggan warns: "The discipline of strategic intuition requires you to recognize when a situation is new and turn off your expert intuition. You must disconnect the old dots and let new ones connect on their own."[3] As safety professionals, we are more susceptible to making snap judgments of situations based on our expert intuition alone.

Strategic insights are moments of clarity that unexpectedly occur. One moment we do not know what to do, the next moment we see clearly what needs to be done, like the Fingersaver example. To improve our strategic thinking skills, we should first recognize our potential for having flashes of insight, and pay closer attention to them when we do.

Though we cannot alter the method by which insights originate, we can enhance their chance of occurrence by knowing how the mind generates a flash of insight. As stated previously, observing things for ourselves helps. It gives us first-hand knowledge. Since personal experience memories are part of strategic intuitions, and are more easily recalled, those observations can contribute to generating insights. We also learned that strategic thinkers dive deeper. Putting more detailed information into our minds gives us more to draw from to generate a flash of insight. Everything we do during the information gathering and analysis phases helps provide more material for our strategic intuition to work.

In strategic thinking, sometimes less is more. In a chapter of the *Handbook of Intuition Research*, which Duggan coauthored with Malia Mason, we find three ways to affect our potential for experiencing strategic intuition. They represent ways we can do more by doing less; less thinking or less focused thinking. They include: *sleep*, *relaxation*, and *defocused attention*.

Sleep

As the adage goes, "Sleep on it." Have you ever woken from a sound sleep with the answer to a question you were pondering? This is an example of strategic intuition. My insight regarding strategic thinking being a process similar to the RCA process was an "Aha!" moment I experienced in the middle of the night.

Duggan and Mason examined the research related to intuition and sleep. Research shows that during sleep, especially the rapid eye movement (REM) stage of sleep, which is the stage rich in dreams, our cognitive associative processes are more flexible. Duggan and Mason state:

> These findings support the view that hyper-associative processes during REM sleep help people draw connections among loosely related material. In addition to promoting fluid thinking, sleep is purported to foster intuition by reshuffling

associations. Evidence suggests that information recombines during REM sleep such that previously overlooked connections are more apparent.[4]

Relaxation

Has anyone ever told you, "Just relax, the answer will come to you." If so, they gave you wise advice. When we focus on a problem, we sometimes arrive at a strategy that seems right at first. We continue to focus on that strategy to improve it, but our efforts fall short. Duggan and Mason warn us:

> The longer the faulty strategy is explicitly considered, the more strongly it becomes associated in memory with the desired outcome. As a result, the problem solver is habitually drawn to consider how the inappropriate strategy can be reworked to produce the solution, despite its ineffectiveness. Although focus is imperative in key stages of intuition–when judging the merit of an insight, for example–it fosters tunnel vision when implemented prematurely. [5]

We often need to take our minds off of a problem we are struggling with in order to solve it. Go think about something else for a while, and you will be surprised how solutions come to mind when you least expect it.

Problem-solving can be stressful when continual thinking on a problem does not deliver a viable solution. Research shows that stress can have a negative impact on the flexibility of cognitive processes. Duggan and Mason summarize the research as follows: "In sum, stress diminishes creative innovation by inducing people to cling to routinized habits, by interfering with effective memory retrieval and by increasing cognitive rigidity."[6] Go relax for a bit. Do something less stressful. Go hang a few pictures.

Defocused Attention

Another piece of advice for fostering strategic intuition is to defocus your attention by spreading it across a wider area. Duggan and Mason state that too much focus can inhibit problem solving, but add that focus is necessary during key phases of the creative process. They add:

> Nevertheless, many researchers speculate that defocused attention in the period leading up to insight fosters novel solutions… and complex problem solving… Evidence suggests that individuals who allocate their attention diffusely or cast broad 'attentional nets' outperform those who focus well, on problems that require the connection of remote problem elements… [7] (Research bibliography citations omitted.)

The takeaway here is to defocus your attention after you have completed the information analysis phase of the strategic thinking process. This will help foster more intuitive and creative solutions.

Creativity

Creativity is the ability to generate new ideas, concepts, or alternatives that are effective in resolving a situation. In the strategic thinking process, creativity helps us generate solutions or strategies.

The question becomes, what do we do if we are not creative by nature? Are there any tools that can help us think more creatively? The answer is yes. We will examine two such methods: Mind Maps and brainstorming.

Mind Maps

One method for enhancing creativity is using the Mind Maps described in the previous chapter. When Tony Buzan created the Mind Map, he included a five-step creative thinking process for using it. In his book, *The Mind Map Book*, he listed those steps:

1. The Quick-Fire Mind Map Burst

2. First Reconstruction and Revision

3. Incubation

4. Second Reconstruction and Revision

5. The Final Stage[8]

The first map lists all thoughts related to the central image or situation, without regard to structure. The second map is a more structured revision that includes the various categories. After the second map, you allow for an incubation period to let the information sink in, just like Duggan and Mason suggested. Give the incubation period a day or more to allow time for your mind's intuitive cognitive mode to generate more insights. After that period, revise the structured map into a more comprehensive final map. Then, you examine the final map for connections between items to help you develop a solution or strategy.

Brainstorming

Another method for enhancing creativity is brainstorming. Before we examine this method further, let us explore a change in the conventional wisdom regarding the left and right brain. When we think of creativity, conventional wisdom suggests that we use the side of our brain where creative thought occurs.

Conventional wisdom is based on the work of Roger Sperry, who in 1981 received a Nobel Prize for his discoveries on the functional specialization of the cerebral hemispheres. According to Sperry, the right side of the brain is the creative or intuitive side, and the left side is the analytical and logical side. Should we apply the two-sided brain theory to improve our creative thinking skills? Not so fast! Let us dive a little deeper first.

In an article titled "How Aha! Really Happens," Duggan cites a study in a 1998 article published in the journal *Neuron*, titled "Cognitive Neuroscience and the Study of Memory." Duggan described the study as providing a more accurate view of creativity. He then wrote:

> Since then, neuroscientists have ceased to accept Sperry's two-sided brain. The new model of the brain is 'intelligent memory,' in which analysis and intuition work together in the mind in all modes of thought. There is no left brain; there is no right. There is only learning and recall, in various combinations, throughout the entire brain.[9]

Although the concept of having two cognitive modes replaced Sperry's two-sided brain model, both theories involve a dominant side or mode. Researchers believe that people have a dominant cognitive mode, either analytical or intuitive. We can use this to our advantage when considering creative thinking techniques.

Duggan wrote about Sperry's discoveries being applied in business: "The most widespread application of Sperry's model was creative brainstorming: People started scheduling meetings where everyone was supposed to turn off their left brains and turn on their right brains, and then let the creative ideas flow."[10] Can you even turn off one side of your brain? Seriously, I cannot. My thinking cap doesn't work that way. We can, however, still benefit from brainstorming with the new concept of *intelligent memory*. Some participants naturally offer analytical viewpoints, while others offer intuitive ones.

Does this mean we should use a brainstorming session to let our creative ideas flow for an effective strategy? Consider this, there is a similar approach that might reverse (hint, hint) how you look at brainstorming.

Traditional brainstorming sessions involve presenting a problem and asking participants to brainstorm solutions. There is an inherent deficiency in that process, though. People rarely devise creative solutions instantly, and these creative solutions are what we seek in this phase of the strategic thinking process. During traditional brainstorming sessions, you get ideas that are not necessarily new or novel.

There is another related technique, however, that is more geared toward creative thinking—*reverse brainstorming*. In his article about Aha! moments, Duggan writes,

> Over the years, companies have used many other techniques that parallel how intelligent memory produces creative ideas. For example, instead of brainstorming to generate creative ideas in an hour or two, some companies do 'reverse brainstorming.' The leaders tell their staff to bring ideas they've had over the past week, for everyone to hear and think about.[11]

This approach reverses the brainstorming process. You might not get as many ideas, but you get better, more creative ideas.

Reverse brainstorming starts with reversing the problem. Consider the hand injury example, and suppose you have completed the information analysis phase. Rather than giving your team this question to ponder before the next meeting, "How can we prevent hand injuries while using hammers and knives?" give them the reverse question, "How can we cause hand injuries while using hammers and knives?" This is actually easier to answer, and can be more fun, too.

Allow some time for the team to think about the reverse problem. This gives their strategic intuition an opportunity to work. When you meet, list those potential negative solutions and rework them into potential positive solutions. Next, brainstorm how to generate a solution or strategy based on the potential positive solutions. Figure 6 provides an example of a simplified brainstorming chart that first reverses the problem. It will help explain the concept.

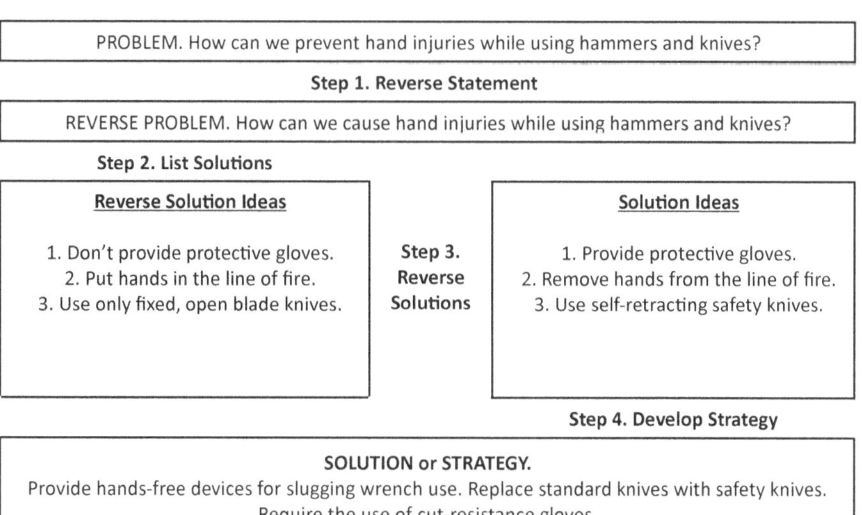

Figure 6.   Reverse Brainstorming Diagram

When it comes time to identify potential solutions or strategies, we are best served by putting our intelligent memory to work for us in finding new and creative ideas. Intelligent memory takes advantage of both our mind's analytical cognitive mode, the logical aspect, and its intuitive cognitive mode, the creative aspect.

Since people have a dominant cognitive mode, it helps to involve a team of people in the process of identifying and evaluating solutions or strategies. This way, you get a better mix of ideas. It is also helpful to involve them in a way that allows time for their strategic intuition to work, to get the best possible creative ideas. Furthermore, when you involve a team, you add the knowledge and experience of others, an ingredient of strategic intuition.

Vision

The last step in the process is creating a strategic vision that will provide the basis for formulating a plan or creating a strategy. Strategic vision is defined by the The Strategy Institute[TM] in their online article titled "Strategic Vision: A Guide for Developing a Clear Roadmap for Your Organization," as follows:

> A strategic vision refers to the desired future state of an organization. It focuses on where the organization aims to be in the long run, usually 3 to 10 years, in terms of its impact, scale, and performance. A strategic vision inspires stakeholders by conveying a compelling and aspirational future while also providing guidance for decision-making, business strategy formulation, and resource allocation.[12]

The key benefit, as it pertains to this chapter, is that it provides guidance for formulating a strategy. In our example, that would be formulating a strategy for reducing hand injuries. Formulating a broader safety strategy is the topic of Part IV.

Creating a strategic vision is the process of imagining and framing plausible future scenarios. All the steps leading up to this point should have helped you clearly define the problem, the critical issue, and identify ways to solve it. If not, review the completed process to determine which steps need further consideration. If so, you are ready to translate all of what you have earned into a vision for the desired future state.

A strategic vision is a type of projection. Projection, in the context of strategic thinking, differs from projection in the psychological sense. Projection in psychology refers to the defense mechanism of attributing to others your own negative traits, fears and insecurities. Subconsciously, it allows you to address an undesired trait without acknowledging it in yourself. An example from relationships is when a person claims that their partner is too controlling, when in fact the person making the claim is the controlling one.

In strategic thinking, projection is more like a movie theater, where a film is being projected onto a screen so others can see it. You are essentially telling a story to others about what the future state can look like. The question is, what story do you want to tell them, what vision of the future do you want them to see? Creating that vision requires some imagination and proper framing.

Imagination

Imagination is another one of those sophisticated, subtle, and at times subconscious hu-man cognitive processes that Mintzberg described. It is the mind's ability to form mental images, generate original ideas, and develop theoretical constructs about the world around us. It improves our cognitive flexibility, creative thinking, and problem-solving.

Curiosity plays a key role in imagination. Think of the imagination of a child compared to that of an adult. Children, having little real world experience compared to adults, are naturally more curious about the world. Their imagination runs wild with creating mental images of how the world could look. As adults, our imagination has been tamed over the years by reality.

Imagination helps you to decide what your projected story or vision is. A simple technique to use is asking yourself, "What-if?" In answering that question, let the sky be the limit. Think more like a child by letting your imagination flow freely. Don't let reality be a killjoy that robs you from seeing what could be.

In the example of hand injuries, information analysis revealed that employees are striking and injuring their hands mostly with hammers and knives. One reverse solution in figure 6 was to put hands in the line of fire. Given that, your question might be, "What if employees never had to place there hands in the line of fire again?" That future scenario would clearly reduce the number of hand injuries in your organization.

Here it helps to distinguish projection from forecasting. Forecasting is using historical data, analysis, and trend continuity to estimate future performance if a change was made. For example, taking historical data from the attribute analysis in table one, we see that three out of seven (~43%) hand injuries occurred while hammering. If we employed the use of hands-free devices for that task, we could estimate a 40% reduction in hand injuries. That is a prediction of the most likely outcome.

A projection, on the other hand, approaches the future in a more imaginative manner, focusing on the long term impact on performance. What if employees never had to place there hands in the line of fire? Conceivably, we could eliminate hand-related injuries altogether.

If that hypothesis held true, what can we do to achieve that vision? What strategic plans would be needed?

The reverse brainstorming example in figure 6 included three actions in the solution statement. Those provide some of the actions needed to achieve that vision. Other actions would be needed given its comprehensive nature.

By using your imagination with the "What if?" technique, you can start to formulate a plan or create a strategy. Once you do, you then must articulate it to the organization.

Framing

Framing is the purposeful use of language to define issues, influence audiences, and guide decision-making toward desired outcomes. It highlights those aspects of the issue that will garner the most support for your strategy, and engagement in the plans to execute it.

In photography and film making, framing is a crucial storytelling tool used to focus audience attention, establish mood, and enhance the narrative. The photographer or director carefully considers the camera's field of view before settling on the final shot. Consider how much better your vacation photographs or videos would be if you had the framing skills of a professional photographer or film maker.

In the hand injury example, you want to focus the organization's attention on the possibility of no longer experiencing hand injuries, and create excitement around helping to achieve that vision. In doing so, you have to convince them that it is at least plausible, if not outright possible. Therein lies the secret of well-framed strategic vision statements, in my opinion, generating excitement around realizing a possibility..

Your strategy statement should state clearly and concisely what you aim to achieve and how you plan to achieve it. You should also include the performance measures.

Below is an example of a strategic vision statement for the hand injury example. You would also need operational and tactical plans for executing the strategy.

Problem: Over 70% of injuries are hand-related. They mainly involve employees striking and injuring their hands while using hammers and knives.

Strategic Vision: Create a work environment and mindset where hands are removed from the line of fire.

Measure: By creating that environment and mindset, we anticipate reducing hand injuries by 40% within a year, and 20% year on year afterwards.

To achieve this, we will [A list of strategic plans.]:

- Develop policies, programs, and procedures for the safe use of hand and power tools.

- Train employees in the safe use of hand and power tools.

- Develop a list of approved cutting tools that identifies the safest tool and the type of glove for each task.

- Provide safety knives for all knife-related cutting tasks to prevent contact with the cutting blade.

- Provide hydraulic torque wrenches to eliminate the need for sledgehammers and slugging wrenches, whenever possible.

- Provide hands-free devices, when needing to use hammers and striking implements, to remove hands from the line of fire or redirect the impacting force.

- Institute a risk assessment and approval process for when safety knives, hydraulic wrenches, or hands-free devices are not workable for the task.

That strategy statement clearly and concisely shows what the objective and desired key results are, including the associated time frame. It then lists the strategic plans designed to achieve that objective. Anyone reading that statement knows what you hope to achieve and how you plan to achieve it.

Given the comprehensive plans associated with the strategy, it conveys the real sense that it is not only plausible, it is achievable. Employees are more likely to get behind a vision that is not only good for them, but is one they can see themselves achieving. You want them thinking to themselves: "We can do this!"

Perhaps the most important, and probably the most obvious aspect of a vision statement is that it has to be future focused. In this case it is. Beyond that, the statement should follow the S.M.A.R.T. Principle, a popular goal-setting framework. That principle states that a goal should meet these requirements:

- S = Specific: The objective must be written so everyone knows what is expected.

- M = Measurable: You need a measure to track progress and determine when the objective is met.

- A = Achievable: You must realistically be able to accomplish the objective.

- R = Relevant: The objective must be important enough to pursue, and should align with the needs of the organization.

- T = Time-bound: You must set a date that is challenging, but not too limiting.

Does that overall vision statement follow the S.M.A.R.T. Principle? You be the judge.

- S = Specific: The objective is clearly defined as a reduction in hand injuries by removing hands from the line of fire.

- M = Measurable: The number of hand injuries can be calculated and measured.

- A = Achievable: A 40% reduction is likely achievable in the time allotted, even though it's challenging, whereas a 100% reduction would likely not be achievable.

- R = Relevant: Hand injuries are the leading body part of all injuries, so reducing them would keep employees safer, and avoid the other negative consequences of hand injuries being prevalent during operations.

- T = Time-bound: The target completion date is the end of the current year, followed by the end of subsequent years.

That vision statement meets the S.M.A.R.T. Principle requirements, in addition to being forward-looking. It is a well-framed statement that will focus the organization's attention on the possibility of no longer experiencing hand injuries, and create excitement around helping to achieve that vision.

By listing the strategic plans under the strategic vision statement, we began formulating a strategy to realize that vision. From that point, you would develop the operational and tactical plans to round out your strategy.

As you can see, the strategic thinking process closely resembles the RCA process. For both, you begin by gathering information, although the mindset is slightly different for the latter. Then you analyze that information to gain insights, versus uncovering causes in an RCA. Finally, you use those insights to create a vision and formulate strategies to realize it, as opposed to planning preventive actions.

I chose the hand injury example to explain strategic thinking because it was a successful strategy that I was deeply involved with during my career. In that joint venture I had divisional safety oversight for, as mentioned in the introduction, hand injuries consistently accounted for about half of our injuries. Reducing hand injuries was a central part of our data-driven strategic approach to injury reduction. Not only did we care about our people, in our industry we had to meet customer requirements regarding injury rates. Therefore, our hand injury reduction strategy was critical to lowering our injury rates, helping to ensure the success of the joint venture.

Our strategic plans involved making a capital expenditure of over $1 million for hydraulic wrenches in our field services operations, equipping every operation with Fingersavers, providing safety knives and cut resistant gloves, and developing approved cutting tool lists. Those plans contributed significantly to our 77% reduction in Total Recordable Injury Rate (TRIR) over five years.

Regarding hand injuries, we all but eliminated those in the years following our strategy. We had no injuries involving hammers and striking implements, and only one injury involving a knife, which involved a new employee. The lesson that taught me was that taking a comprehensive approach to solving the problem once and for all was the right approach. It you are going to do something, do it right the first time!

In this and the previous chapter, we examined the three phases of the strategic thinking process. We examined the various elements of each phase of the process, using an example from a safety perspective.

In Part III, we will cover strategic analysis for each of the four Strategic Areas of Safety. In doing so, we will present various techniques for performing more formal analyses in those areas.

Endnotes

1. William Duggan, *Strategic Intuition*, (New York: Columbia Business School Publishing, 2007), 1.

2. Ibid., 2.

3. Ibid..

4. William Duggan, Malia Mason, "Strategic Intuition," chap. 7 in *Handbook of Intuition Research*, ed. Marta Sinclair (Northampton: Edward Edgar Publishing, 2011), 80-81, accessed January 22, 2022, https://business.columbia.edu/faculty/research/strategic-intuition.

5. Ibid., 81-82.

6. Ibid., 82.

7. Ibid., 83.

8. Tony Buzan, *The Mind Map Book* (New York: Dutton, 1993), 156-161.

9. William Duggan, "How Aha Really Happens," *strategy+business,* November 23, 2010, accessed January 22, 2022, https://www.strategy-business.com/article/10405.

10. Ibid.

11. Ibid.

12. The Strategy Institutet™, "Strategic Vision: A Guide for Developing a Clear Roadmap for Your Organization" *The Strategy Institute*, September 11, 2024 accessed April 8, 2026, https://www.thestrategyinstitute.org/insights/strategic-vision-a-guide-for-developing-a-clear-roadmap-for-your-organization

PART III
Strategic Analysis

Chapter Six:
Performance Analysis

Hence the saying: If you know the enemy and know yourself, you need not fear the result of a hundred battles. — Sun Tzŭ

Ohmae said that analysis is the starting point of strategic thinking. The same holds true for strategic planning; it begins with analysis. Analysis is a detailed examination of something complex in order to understand its parts. The purpose in analyzing your safety program is to give you a clearer understanding of its current state and where it needs to be improved.

The performance strategic area focuses on activities associated with eliminating or mitigating risks, and the measure of their success. Analyzing this area helps answer the questions: "How safe are we, how safe do we want to be, and how can we be safer?"

One of our primary objectives as safety professionals is to reduce risk. Risk is the likelihood of a hazard, a potential source of harm, actually causing harm. Harmful events occur when causal factors weaken our defenses, the barriers we erect to protect against the harmful effects of hazards. Therefore, we can say that hazards and causal factors are enemies that negatively affect our safety performance. Bearing this in mind, we can learn from the strategies of war.

In *The Art of War*, Sun Tzŭ wrote:

Hence the saying: If you know the enemy and know yourself, you need not fear the result of a hundred battles. If you know yourself but not the enemy, for every victory gained you will also suffer a defeat. If you know neither the enemy nor yourself, you will succumb in every battle.[1]

How well do you know the risks associated with your operations? Strategic analysis will help you know them better. Chapter Three introduced the following two assessments for the performance strategic area that will help you identify areas of risk:

1. *Risk Profile:* The first and foundational assessment is a Risk Profile or Risk Registry. A Risk Profile helps you to know your enemy, hazards, the potential source of harm. It is a prospective assessment. It looks at harmful events you could experience in the future, and the risk of experiencing those events.

2. *Attribute Analysis:* The second assessment is an Attribute Analysis. An Attribute Analysis helps you to know your other enemy, causal factors, the factors that weaken your defenses. It is a retrospective assessment. It looks at the harmful events you experienced in the past, and the reasons you experienced them.

These two assessments offer an excellent source of insight into what objectives to include in your safety strategy. This chapter introduces these assessments and presents a practical approach for performing them. The primary objective is to help you see the value in performing them as part of your strategic planning process.

Risk Profile

What is a Risk Profile? It is a document that identifies all the potential safety risks associated with your company's facilities and operations. Since risk is the likelihood of a hazard causing harm, a Risk Profile starts with identifying potential hazards. It then assesses the risks associated with them and identifies the control measures needed to eliminate those risks or mitigate them to an acceptable level.

If your company is certified in, or pursuing certification in, one of the International Organization for Standardization's (ISO) management system standards, you know a Risk Profile by another name, as described below. Both processes are what this book refers to as a Risk Profile.

1. *ISO 45001* - "Occupational health and safety management systems—Requirements with guidance for use," Clause 6.1.2, "Hazard identification and assessment of OH&S risks," requires companies to identify all of their safety and health hazards and assess their associated risks.

2. *ISO 14001* - "Environmental management systems—Requirements with guidance for use," Clause 6.1.2, "Environmental aspects," requires companies to identify all of their environmental aspects and assess their potential impacts.

A Risk Profile is something you should have regardless of any strategic planning process. The extra benefit it provides to strategic planning is that it identifies risks that have not been adequately mitigated. It helps to identify issues that could be detrimental to the success of your safety program in the future. Since the Risk Profile is dynamic, due to changes in facilities and operations, reviewing and updating it should be a regular part of your annual strategic planning process.

The process works best when it includes a scoring system to generate a numerical risk ranking to compare the levels of risk and prioritize mitigation efforts. A scoring system suitable for this purpose is found in using the Failure Modes and Effects Analysis (FMEA) method.

If your company is involved with military, process, or manufacturing systems, you might recognize that a Risk Profile is much like an FMEA. An FMEA is a method used for identifying and evaluating potential failures in a design, a process, or a product. Developed by the military in the 1940s, it has become one of the more popular risk management tools and is used in a variety of applications beyond what it was created for. We will use it here to create a Risk Profile. Figure 7 shows the structure of a basic FMEA spreadsheet with the header rows in split-view.

System, Area, or Job	Hazard	Failure Event	Potential Effects	SEV	Potential Causes	Existing Controls	OCC	Detection Methods	DET	RPN
System, Area, or Job	Hazard	Mitigating Actions		SEV	OCC	DET	RPN	Risk Reduction		

Figure 7.   Risk Profile - FMEA

Developing a Risk Profile using the FMEA method for strategic planning includes three basic steps: *identification*, *assessment*, and *mitigation*, as described below.

Preparation

To perform an FMEA, you first need to develop an FMEA spreadsheet with the columns shown in figure 7. The next tool you need is a scoring matrix. The FMEA method uses three scores: severity (SEV), occurrence (OCC), and detection (DET). These scores are multiplied together to generate a Risk Priority Number (RPN).

(Eq. 1)   *RPN = Severity (SEV) x Occurrence (OCC) x Detection (DET).*

As the RPN value increases, so does the risk. You can perform an FMEA using a ten-point (1-10) low to high scale, which is recommended because it provides more granularity in RPNs. This makes it easier to prioritize actions. For demonstration purposes, the example below uses a five-point (1-5) low to high score, similar to the basic 5 x 5 Risk Matrix. Table 2 shows an FMEA scoring matrix for severity.

Table 2.   FMEA Scoring Matrix - Severity

5	Very High (Catastrophic)	Death or multiple serious injuries. Major external environmental impact. Major financial or operational impact.
4	High (Critical)	Permanent disabling injury. Significant external environmental Impact. Significant financial or operational impact.
3	Moderate (Serious)	Injury with temporary disability. Serious internal environmental impact. Moderate financial or operational impact.
2	Low (Minor)	Medical treatment only case. Minimal internal environmental impact. Minor financial or operational impact.
1	Very Low (Negligible)	First aid case or no injury. Negligible internal environmental impact. Negligible financial or operational impact.

Table 3 shows FMEA scoring matrices for occurrence and detection. The detection scale is the opposite of the severity and occurrence scales. Detection is the likelihood of an event being detected before it occurs. A detection score of five or ten is the lowest likelihood of the event being detected, whereas a one is the highest likelihood, representing the lowest risk.

Table 3.   FMEA Scoring Matrices – Occurrence & Detection

5	Very Likely (Frequent)	Occurs at least monthly.
4	Likely (Often)	Occurs more than once a year.
3	Probable (Occasional)	Occurs about once a year.
2	Unlikely (Seldom)	Has or could occur once every 5 years.
1	Very Unlikely (Rarely)	Has not occurred in more than 5 years.

1	Very High	Controls will most likely detect a pending failure.
2	High	Controls have a good chance of detecting a pending failure..
3	Moderate	Controls may detect a pending failure.
4	Low	Controls have a poor chance of detecting a pending failure.
5	Very Low	Controls probably will not detect a pending failure.

Another tool needed is an acceptability matrix. Organizations often develop their own acceptability matrix, based on their risk tolerance. Table 4 shows the levels of acceptability used in this example.

Table 4.   FMEA Acceptability Matrix

RPN	Acceptability	Action
1 to 14	Acceptable	Risk is low, or as low as reasonably practicable. Action is not required unless the risk level changes..
15 to 29	Tolerable	Risk could be further mitigated. If action is taken, the timeline for completion should be reasonable .
30 to 49	Undesirable	Risk is higher than desired and actions needs to be taken at the earliest convenient opportunity to reduce the risk to a tolerable or acceptable level.
Above 49	Unacceptable	Risk is too high and immediate action is required to reduce the risk to a tolerable or acceptable level.

The FMEA process is best performed by a small team knowledgeable in the operations associated with your business. A team approach helps ensure all hazards are identified and makes the scoring process less subjective.

Step 1 - Identification

The identification step is best completed through brainstorming, since the objective is to list as many hazards as possible. You should start with the system, area, or job column first, and list all of those before identifying the hazards in each. Obviously, each system, area, or job could have multiple hazards, so your list will fill out rather quickly.

Throughout your brainstorming session, you will most likely bounce between areas to capture thoughts that come to mind. Take the time to go back and add those to your list so they don't get forgotten.

It is important during this step to capture as many hazards as you can without considering their current level of risk or existing controls and methods of detection. These aspects of the hazards will get addressed later in the process. Consider all types of hazards, including, but not limited to:

- General

- Physical

- Chemical

- Biological

- Ergonomic

Step 2 - Assessment

This step considers five related aspects of the hazard, as described below.

1. *Failure Event:* Potential failure(s) that might happen, related to the hazard, which could cause a harmful event.

2. *Potential Effects:* The potential effect(s) of a failure event if it occurred. A severity score is assigned based on the potential effects. A severity score of one is the lowest severity, and a five or ten is the highest severity.

3. *Potential Causes:* Potential cause(s) of the failure event.

4. *Existing Controls:* The controls currently in place designed to prevent the failure event from occurring. An occurrence score is assigned based on those controls. Occurrence is the likelihood of an event happening given the existing controls. An occurrence score of one is the lowest likelihood of occurrence, and a five or ten is the highest.

5. *Detection Methods:* Existing controls that would provide early detection or warning of a potential failure event. A detection score is assigned based on those controls. Detection is the likelihood of an event being detected before it occurs. As previously stated, a detection score of five or ten is the lowest likelihood of the event being detected, whereas a one is the highest likelihood. The scale is simply inverted.

When scoring these categories, the five-point scale makes it easier to arrive at a consensus over each score. A ten-point scoring system makes consensus more difficult to achieve, but reduces the chance of having duplicate RPNs, which would make prioritization more challenging. Do not be overly concerned with being as accurate as possible with scoring. The whole process is somewhat subjective anyway. The RPN is the key value. It is more important to be consistent during your scoring efforts than to be pinpoint accurate.

Table 5 shows example results of the assessment for two of the hazards presented above. This step is the heart of the FMEA process, and the objective is to generate an RPN so that hazards can be evaluated for acceptability and prioritized for action.

Table 5.   FMEA Assessment

Identification		Assessment								
System Area, or Job	Hazard	Failure Event	Potential Effects	S E V	Potential Causes	Existing Controls	O C C	Detection Methods	D E T	R P N
Lifting Turbine Casings & Rotors	Heavy Component Falling	Defective rigging	Injuries, property damage, project delays	4	Failure to inspect, old equipment	QRLY Rigging inspection	2	Preuse rigging inspection	2	16
		Improper Rigging		4	Lack of rigging knowledge or skills	Initial Training	3	Supervisor observation during lift	3	36
GSU XFMR Power Factor (Doble) Testing	Electro-cution	Tech. contacts test voltage	Death or serious injury	5	Distracted tech. not following procedure	Safety observer with safety switch	2	Supervisor oversight during test	3	30
		GSU XFMR circuit not isolated		5	Failure to properly isolate system	Lockout Program	2	Pretest safety checks	1	10

The RPN column illustrates how it provides a ranked list of risks. Upon completing the FMEA, you have a profile of the risks to your organization presented by the hazards associated with your facility and operations.

Step 3 - Mitigation

Table 6 shows the last step of the FMEA process. This step documents the actions planned to reduce the risk for each hazard requiring mitigation. In this example, the mitigation step was only applied to hazards having unacceptable or undesirable risk levels, which need more or better controls to lower the risk to an acceptable level.

Table 6.   FMEA Mitigation

Assessment		Mitigation					
Failure Event	R P N	Mitigating Actions	S E V	O C C	D E T	R P N	Risk Reduction
Defective Rigging	16	TOLERABLE - NAR					
Improper Rigging	36	UNDESIRABLE - Annual Rigging & Lifting Training and qualification	4	2	2	16	-56%
Tech. contacts test voltage	30	UNDESIRABLE - Use only Doble certified technicians	5	2	2	20	-33%
GSU XFMR circuit not isolated	10	ACCEPTABLE - NAR					

Using the acceptability matrix, the risk with an RPN of ten would be acceptable. The risk with an RPN of sixteen would be tolerable. Neither require action as noted by the NAR (No Action Required). The other two risks are undesirable, based on their RPN scores, and warrant establishing mitigating actions.

After determining the actions needed to mitigate the risk, assign new severity, occurrence, and detection scores, based on the mitigation, and calculate a new RPN. Afterward, compare the old and new RPNs to determine the percentage of risk reduction using the percent change equation below.

*(Eq. 2)   Percent Change = (New Value - Old Value) / Old Value * 100*

You will see in tables 5 and 6 that there is no difference between the old and new severity scores for the undesirable risks. Seldom will you be able to take action to reduce the severity, although it is theoretically possible in some cases. The vast majority of times, however, you will only be able to improve the occurrence and/or detection scores.

The risk reduction value provides you with a metric that quantifies the reduction for each applicable line item. You can also apply it to the Risk Profile overall. The results provide a useful metric for your strategic plan.

Strategy

Having completed the FMEA, it is time to consider what to incorporate into your strategic planning process. Based on the example, you might include a combined objective such as: "Reduce the potential risk of high severity hazards identified in a Risk Profile, as measured by a 40% reduction in the combined Risk Priority Number." It would include operational and tactical plans related to the mitigating actions stated in the profile.

Notice how the potential overall risk reduction percentage was used to quantify the measure of success. To achieve that target, multiple mitigating actions may need to be taken, but at least you know before executing your strategy that success is achievable. A 40% reduction in risk is realistic and achievable, though challenging. As *The Art of War* teaches, ensure victory before going into battle. Knowing that your risk reduction is realistic and achievable helps to confirm that you have a winnable battle plan.

On a final note, if you are creating a Risk Profile for ISO certification, avoid treating it like a check the box exercise. My impression after reviewing multiple Risk Profiles used for ISO certifications is that some were completed only to the degree necessary to pass a certification audit. They met the intent of the standard, not the spirit of it. As a result, they were nominally effective in reducing risk, and left a lot of hazards unaccounted for.

Attribute Analysis

What is an Attribute Analysis? An attribute is a characteristic of something. An Attribute Analysis is the examination of the characteristics associated with records in a data set. The key benefit is that it allows you to extract usable information from a data set, also known as *data mining*. How usable the information is will depend on how good your data is.

A real-world example is the study of Serious Injuries and Fatalities (SIFs) that was performed within the safety profession. In February 2017, I attended a luncheon at the (then) *ASSE SeminarFest* in Las Vegas. Tom Krause, Ph.D., founder of Behavioral Science Technology, presented a talk titled "The Future of Behavior-Based Safety." His talk mentioned the study of SIFs. That subject intrigued me due to the potential for serious injuries and fatalities in our industry, and prompted me to research the study further.

What I learned about the injury data behind the study made my research worthwhile. The SIF study included injury data for 2008 and 2009 provided by six companies from different industries.

In their September 2015 *Professional Safety* article titled "Preventing Serious Injuries & Fatalities: Study Reveals Precursors & Paradigms," Donald K. Martin and Alison A. Black had this to say about the data behind the study:

> While all participants had comprehensive incident investigation systems and reporting structures, the maturity and information contained in the individual databases varied greatly. The least sophisticated systems contained little more than unique incident identification numbers, indications of actual consequence and incident narratives.[2]

The quality of data used for an Attribute Analysis is important. Poor-quality data does not necessarily prevent you from mining it to uncover areas needing further examination, but good-quality data makes it much easier and produces far better results.

More recently, Dr. Krause commented on the study from another learning perspective in his blog post titled "Preventing Serious Injuries & Fatalities (SIFs) Requires Organizational Learning." He wrote:

> We confirmed that studying incidents as a series yields understandings that are not apparent in single events. We also found that in order to study a series effectively

you must have good incident reports. Understanding the first point leads to the importance of the second one.

1. Learning effectively from incidents requires looking at a series, at least thirty.

2. In order to be useful, incident reports must explain what actually happened that resulted in the injury.[3]

Dr. Krause went on: "For a variety of reasons root cause analysis of single events, doesn't usually get to actionable data...But a good set of incident reports analyzed as a series can yield very useful information."[4]

The takeaway from these comments is twofold. First, studying incidents as a series, like you would during an Attribute Analysis, can reveal things you might not see when studying a single event, such as during a Root Cause Analysis (RCA). Second, at least thirty events are needed to effectively learn from the analysis of the series.

When trying to understand the value of an Attribute Analysis, it helps to compare it to an RCA. An RCA examines one incident, whereas an Attribute Analysis examines a group of incidents. An RCA aims to prevent future incidents similar to the one examined, whereas an Attribute Analysis aims to prevent future incidents that share similar characteristics as those found to be most prevalent. The difference between the two may seem subtle, but it is important.

Now we will review the key aspects of an Attribute Analysis, including: *attributes*, *attribute data*, *attribute selection*, and *data mining*. An example is provided to show its usefulness.

Attributes

Attributes are characteristics of something; in this case, safety incidents. Different types of safety incidents, such as injury events, process safety events, fires and explosions, vehicle accidents, or environmental releases, have some attributes in common and some attributes that are unique. This chapter will use injury events to explain the process. You would use the same process for other safety-related incidents.

Table 7 provides an example of attributes of injuries that can be analyzed to glean valuable information for strategic planning. The event and employee attributes are transferable to other types of incidents, as you can see by the listed attributes.

Table 7.   Injury Attributes

Event Attributes	Employee Attributes	Injury Attributes	Causal Attributes
Incident Number	Age	Body Part	Event/Exposure (Hazard)
Incident Date/Day of Week	Discipline/Role	Injury or Illness Type	Source of Injury
Incident Time/Shift	Time in Role	Severity Level	Tool or Equipment Involved
Division or Region	Hours at Work	Classification (FAC, MTC, etc.)	Substance Involved
Business Unit or Facility	Training in Topic	Days Restricted	Immediate Cause
Shop/Area/Site Location	Past Related Injuries	Days Lost	Root Cause

An Attribute Analysis is a retrospective assessment. It looks back at events that have already occurred and helps reveal associated weaknesses. Suppose your business unit has a sizable population of older employees and you want to know if the aging workforce accounted for a disproportionate number of injuries. Human Resources informs you that 23% of employees are between the ages of fifty-five and sixty-four. If 32% of injuries occurred in that age group, it would be a disproportionate number of injuries. That would be an area of weakness. In that case, you would consider ways to address safety concerns with an aging workforce in your safety strategy.

Attribute Data

The first order of business in performing an Attribute Analysis is gathering the data. Hopefully, your organization uses incident reporting software, which is going to be the primary source of your data. If not, go old school and create your own database or spreadsheet and populate it with data from your individual injury reports.

Your reporting software might already have all the attributes you need. If not, you will need to download the raw data from that application in spreadsheet format and add other attributes to it, thus providing a secondary source of data. You might also find that the list of selections under some attributes used by your organization are not suitable. If so, you will also need to enter that data into your own spreadsheet.

Incident reporting software typically provides tools to do an analysis with. Having used several incident reporting applications, my experience is that these analytical tools provide the basics. For example, you can only select one attribute at a time to analyze, such as body part. The applications often provide a pie chart showing the number and percentage of injuries for each body part. That might get you started, but further analysis will be required to help you understand why a particular body part keeps getting injured. I explain below how to cascade attributes to get more actionable data.

An important consideration before beginning your Attribute Analysis is the size of your data set. As Dr. Krause revealed, you need a series of at least thirty incidents.

At the business unit level, you might have only a few recordable injuries a year, or have gone several years without one. If you do not have a large enough series, you can do a few things to expand it.

First, including first aid cases along with recordable injuries in your analysis will increase the size of your data set. Severity of consequences is the only substantial difference between first aid cases and injury events. The last time a specific event occurred may have resulted in a first aid case. However, it is possible that it could lead to a more serious injury in the future. The only caveat is that you must include the same attributes as you do for recordable injuries so that your data is comparable.

Second, you can add injuries from other business units in your company, if that data is available to you, provided they have the same type of operations. Injuries resulting during their operations could also occur during your operations. If that added data set has the same attributes as yours, it is an effective way to increase the size of your data set.

Another concern regarding your data set is its age. How far back should you go? OSHA record-keeping rules require preserving injury records for five years. That makes for a reasonable maximum time frame. Going back even farther might not produce results representative of current operations. Or, your company might have addressed an injury trend six years ago, and going back ten years would flag that trend again as needing correction.

A minimum time frame is going back far enough to capture the prior calendar year's data, or at least the previous twelve months. This way you capture any seasonally dependent injuries, as well as the increased risk of injury during peak operating cycles

Another concern regarding your data set is its quality. As the G.I.G.O. Principle says, "Garbage In, Garbage Out." Three common data quality issues found in injury event records are: *incompleteness*, *inconsistency*, and *noise*.

1. *Incompleteness:* This refers to missing information in the data set, which is often found in attributes that are optional as opposed to required fields in the reporting process. Two common examples include employee age and time in role, both of which can provide valuable information. When presenting Attribute Analysis results to leadership, it is no fun to add the disclaimer that some results are inconclusive because a large percentage of records are missing data.

2. *Inconsistency:* This refers to variability in the data set. Inconsistent data is especially common in reporting processes where multiple people are making the entries. Some attributes are more subjective to inconsistency than other attri-

butes. Immediate or root causes is one example where people have different ways of looking at incident causes. Inconsistencies are also common in attributes having a long list of selections.

3. *Noise:* This refers to random or irrelevant data in the data set. Noise is common in injury data where attributes have a long list of selections, especially where multiple selections can be made. It is my experience that hazard and cause-related attributes are the primary culprits. Reporting data often includes every potential hazard or cause, for a given injury, with multiple items selected. That leaves the data useless from an analysis perspective. The old idiom "less is more" applies in this case. Having just the predominant hazard or cause selected is better. The more attributes that are selected, the harder it is to identify a meaningful group.

Attribute Selections

It is important to have meaningful selections for each attribute. Table 8 provides lists of injury attributes used by the Bureau of Labor Statistics (BLS) in one of its Injuries, Illnesses, and Fatalities (IIF) data sets.

Using the BLS attributes allows for benchmarking your organization against others in your industry classification, though unlike injury rates it will require some effort to develop comparison data. For example, to compare the percentage of injuries in a given age group you will have to calculate that yourself.

Interestingly, the BLS 2021-2022 biennial data shows only 17% of injuries from all industries involved employees in the 55 to 64 age group. That is lower than those in the 25 to 34 (22%), 35 to 44 (20%), and 25 to 54 (19%) age groups. Younger employees tend to have slightly more injuries.

There are five basic types of data in a data set. Below is a description of these data types.

1. *Binary Data:* Binary data has only two values. True/false or yes/no responses are binary. Examples of binary data include training in a topic (did the employee receive it) or past related injuries (did the employee or company have any.)

2. *Interval Data:* This data type represents quantitative data with equal intervals between consecutive values. Examples of interval data include date and time of incident. Interval attributes do not require a list of selections. Day of the week or shift represents a slightly different type of data that is more useful than date and time. You can easily convert dates into weekdays using a spread-

sheet formula. You can convert time to a work shift using a lookup table based on your shift schedules. The amount of time before or after breaks, or before the end of shift, that an injury occurred can also be valuable information, due to changes in an employee's attention throughout the shift.

3. *Ordinal Data:* This data type represents qualitative data that can be ranked in a particular order. Examples of ordinal data include age and severity. The BLS age group attribute shows that this type of data requires a list of selections, albeit, a relatively short list. Your company's reporting software likely already contains a selection for severity, and possibly even potential. If not, you can use a similar severity scale as that used in your 5 x 5 Risk Matrix and add it to a secondary data set.

Table 8.   BLS Attributes

Sprains, strains, tears	Head	Contact with object, equipment	Fires and explosions
Fractures	Eye	Falls, slips, trips	Exposure to harmful substances or environments
Cuts, lacerations, punctures	Neck	Slips, trips without fall	All other events or exposures
Bruises, contusions	Trunk	Fall to lower level	
Heat (thermal) burns	Back	Fall on same level	14 to 15
Chemical burns and corrosions	Shoulder	Overexertion and bodily reaction	16 to 19
Amputations	Arm	Overexertion in lifting or lowering	20 to 24
Carpal tunnel syndrome	Hand	Repetitive motion involving micro tasks	25 to 34
Tendinitis	Wrist	Struck by object or equipment	35 to 44
Multiple traumatic injuries	Knee	Struck against object or equipment	45 to 54
Soreness, pain	Ankle	Caught in or compressed by object or equipment	55 to 64
All other injuries	Foot	Violence and other injuries by persons or animal	65 and over
	Body systems	Intentional injury by other person	
Skin diseases or disorders	Multiple body parts	Injury by person-unintentional or intent unknown	
Respiratory conditions	All other body parts	Animal and insect related incidents	
Poisonings			
Hearing loss			
All other illnesses			

4. *Nominal Data:* This data type represents qualitative data that cannot be ranked in order or hierarchy. Examples of nominal data include body part, injury or illness type, OSHA classification, among others. Some nominal attributes have a finite list of selections available, such as body part. Others do not, or they can have an unreasonable number of selections. One example is event or exposure type, or what some companies call hazards. This can cause noise as described above.

Although the list of selections in the BLS event or exposure attribute in table 8 seems unreasonably long, sometimes the BLS incorporates a two-part attribute using both categories and types. This approach can also be overdone, still ending up with noise.

Another solution seen in the BLS attributes is the "All Other" categories. Since you will only find small percentages of injuries having attributes other than those listed, there is no concern regarding missing valuable data.

5. *Textual Data:* This data type represents unstructured data as text. One example is the incident description. While not categorical, it is still valuable information during an Attribute Analysis. This is especially true if you have to add your own attribute data based on what you learn from the incident description.

One drawback of text data is that it is often unstructured and time-consuming to read. To address this, some companies use a standard format or template for creating a standard incident description, making key information easier to find in the text. The format requires placing key information in a set order within the description.

Data Mining

Data mining can go to multiple depths. An example of single-level data mining related to an aging workforce was given above. Performing the same level of data mining for all attributes is a good first step to help you focus your efforts further.

Table 9 includes the five most valuable attributes for performing an injury Attribute Analysis, in my opinion. It shows the top five most prevalent categories under each attribute. The data set used contains one hundred hypothetical records representing injuries that occurred over three years. (NOTE: Since the count is one hundred, it serve as percentages as well.)

Table 9.   Top 5 Injuries

Hand & Fingers	37	Cuts, Lacerations & Grazes	31	Struck With or Against	27	Parts, Materials & Containers	27	Work Execution (Awareness)	44
Ankle, Knee & Leg	13	Sprains, Strains & Twisted	30	Slip, Trip or Fall	23	Hand & Power Tools	20	Manual Handling (Ergonomics)	18
Back & Spine	8	Fractures & Dislocations	10	Physical (Body Mechanics)	18	Shop & Portable Machinery	12	Housekeeping & Upkeep	10
Arm & Wrist	7	Bruised or Swollen	7	Struck By Flying or Falling Objects	14	Ladders & Stairs	7	Plant & Equipment Design	6
Eyes	7	Foreign Bodies in Eye or Eye Irritation	7	Caught In or Between	4	Other Exposures or Sources	6	Proper PPE	6

Showing the top five categories under each attribute is a single-level analysis. Each column can be portrayed in a simple pie chart. That is a good start, but stopping at that level yields information that is often too general to base a targeted strategy around.

Table 10 represents the cascading feature I built into my spreadsheet, which allowed for diving deeper for more meaningful information. I initially created it the hard way with tables, formulas, and macros. Using Microsoft Excel pivot tables and slicers, with a nice custom dashboard, would have made it more streamlined and easier to use.

Table 10.   Hand & Finger Injuries

Hand & Fingers	37	Cuts, Lacerations & Grazes	22	Struck With or Against	15	Parts, Materials & Containers	6	Work Execution (Awareness)	5
						Hand & Power Tools	5	Proper PPE	1
						Shop & Portable Machinery	3	Compliance Failure	0
						Mobile Equipment	1	Environmental Incident	0
						Cords & Hoses	0	Equipment Safety	0

Column A, Body Part, illustrates hand or finger injuries represented 37% of all injuries. Choosing only that body part restricted the following categories to only those injuries involving hands or fingers. The same restriction applied to subsequent columns. (NOTE: The numbers or percentages in the "#/%" columns are still percent of all findings.)

The analysis revealed that cuts or lacerations to hands or fingers comprised 22% of injury cases, and 15% were due to of being struck by or against something. Most often parts and materials or hand and power tools caused the cut or laceration. That is significantly more detailed information than simply knowing hand or finger injuries represent 37% of all injuries. Diving even deeper to find which parts, materials, or tools were involved would give you even more detailed information for developing tactical plans. You might consider performing a Job Safety Analysis (JSA) of related jobs to identify areas for improvement. If you do not have a JHA process, you might consider adding one.

Table 11 shows the cascading analysis, without the body part, to examine all cuts and lacerations. It reveals similar results to the examination of hand or finger injuries, so no new information was discovered. This happens during data mining. You do not always strike gold. When you do, the nuggets you uncover can be invaluable.

Table 11.   Cuts & Laceration Injuries

Injury Type	#/%	Event Type	#/%	Source Type	#/%	Causal Theme	#/%
Cuts, Lacerations & Grazes	31	Struck With or Against	17	Hand & Power Tools	6	Tool Safety	3
				Parts, Materials & Containers	6	Work Execution (Awareness)	2
				Shop & Portable Machinery	3	Manual Handling (Ergonomics)	1
				Cranes, Rigging & Lifting Equipment	1	Compliance Failure	0
				Mobile Equipment	1	Environmental Incident	0

Table 12 shows the cascading analysis of the second most prevalent injury type, sprains and strains. Here we discover that 18% of all injuries are sprains or strains due to poor body mechanics. This occurs most often while handling parts and materials. You could perform ergonomic assessments on material handling jobs to dig deeper into the exact cause. If you do not currently perform ergonomic assessments, you should consider doing so as part of your safety strategy.

Table 12.   Sprain & Strain Injuries

		Physical (Body Mechanics)	18	Parts, Materials & Containers	8	Manual Handling (Ergonomics)	7
Sprains, Strains & Twisted	30			Plant Systems & Equipment	3	Plant & Equipment Design	1
				Shop & Portable Machinery	3	Compliance Failure	0
				Cranes, Rigging & Lifting Equipment	1	Environmental Incident	0
				Floors, Walkways & Other Surfaces	1	Equipment Safety	0

Table 13 shows the cascading analysis in the reverse direction to show the breakdown of body parts for the top two injury types. It shows another way of analyzing the data.

Table 13.   Injury Type & Body Part

Cuts, Lacerations & Grazes	31	Hand & Fingers	22		Sprains, Strains & Twisted	30	Ankle, Knee & Leg	7
		Ankle, Knee & Leg	3				Back & Spine	6
		Head or Neck	3				Shoulder or Upper Arm	6
		Arm & Wrist	2				Head or Neck	3
		Face or Jaw	1				Arm & Wrist	2

Table 14 shows select attributes in the reverse order. It highlights the causal theme of work execution (awareness), which was the top attribute category, representing 44% of all injuries. That causal theme represented events where human factors was a predominant causal factor, including stress, fatigue, distraction, rushing, or other reasons the employee did not pay close enough attention to the task at hand.

Table 14.   Work Execution

Work Execution (Awareness)	44	Cuts, Lacerations & Grazes	20	Hand & Fingers	15
		Sprains, Strains & Twisted	7	Ankle, Knee & Leg	2
		Crushing	4	Arm & Wrist	1
		Fractures & Dislocations	4	Face or Jaw	1
		Bite/Sting	3	Head or Neck	1

When cascading through the attributes in reverse order, it was helpful not to include specific attributes, such as event and source type, to reveal that 20% of all injuries resulted in cuts and lacerations because of the lack of awareness while work was being done. As I see it, that knowledge adds to the shock value of the information. Telling employees that if they want to increase their likelihood of getting cut, something no employee would want to do on purpose, all they need to do is get distracted while working. Hopefully, that alone would cause them to pay closer attention.

The event and source types could be added to the end of that analysis to get more detail on how the lacerations are occurring, which would be helpful for training purposes. That is one of the benefits of a well-designed Attribute Analysis tool, it allows you to slice and dice the information in multiple ways.

This cascading exercise shows how further examination of a single attribute helps to narrow your focus, pointing you to the right target. It takes this deeper examination to get to this actionable information.

Figure 8 helps to illustrate how cascading attributes get you closer to the bullseye. As you add more attributes into the analysis, the more defined the group of records becomes. You go from a general problem area to a specific one.

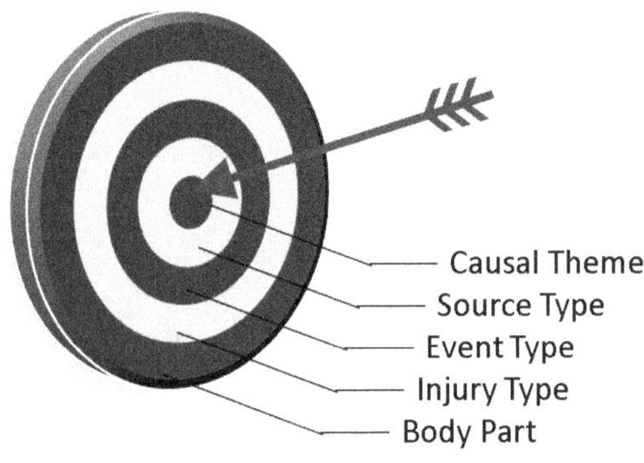

Figure 8.   Targeted Injury Reduction

This reminds me of Zig Ziglar, one of my favorite motivational speakers when I was a young man. One memorable quote is: "Don't become a wandering generality. Be a meaningful specific."[5] When it comes to reducing incidents, such as injuries, it is far better to drill down to the specifics.

This exercise should have shown how an Attribute Analysis can provide insight into your company's safety performance that the investigation and analysis of a single incident event might not reveal. If your organization's RCA process failed to drive needed safety improvements in the past, an Attribute Analysis can pick up the slack and help you address these improvements in your safety strategy.

Where a RCA may have identified unique contributing and causal factors for each incident, the Attribute analysis can bring to light more common denominators for all incidents. Both of these analysis techniques provide valuable information for preventing incidents, but each has its own way of adding value, helping you to learn more about the potential risks from different perspectives.

Whether your vision is achieving safety excellence, zero injuries, or something similar, understanding your safety performance is critical to realizing that vision. You must know what future events could affect your performance, the risk of them occurring, and how to reduce that risk. This is what a Risk Profile is for. It is also important to know what past events have negatively affected your safety performance, and to take the necessary steps to prevent them from recurring. This is what an Attribute Analysis is for.

The next chapter examines the analysis of compliance, the second strategic area of safety. This analysis will include methods similar to those used in analyzing performance, namely the use of profiles and an Attribute Analysis.

Endnotes

1. Lionel Giles, *Sun Tzŭ on the Art of War*. [1910; Project Gutenberg, 2004], Chapter IIII, Paragraph 18, https://www.gutenberg.org/files/132/132-h/132-h.htm.

2. Donald K. Martin and Alison A. Black, "Preventing Serious Injuries & Fatalities: Study Reveals Precursors & Paradigms." *Professional Safety,* September 2015, 36-37.

3. Tom Krause, "Preventing Serious Injuries & Fatalities (SIFs) Requires Organizational Learning," *Krause Bell Group*, February 26, 2019, accessed October 21, 2023, https://krausebellgroup.com/learning-to-prevent-serious-injuries-and-fatalities-sifs/.

4. Ibid.

5. Zig Ziglar, "Meaningful Specific," *Ziglar*, accessed October 22, 2023, https://www.ziglar.com/quotes/meaningful-specific/.

Chapter Seven:
Compliance Analysis

In effect, the research and theory say that your audit program must look at the things that are right for your organization. — Dan Peterson

The advent of 45 rpm records unleashed the explosion of rock-n-roll music in the 1950s. It also ushered in the golden age of jukeboxes. The hit songs of the day were on the A-side of the record, but there were also songs on the B-side. The A-side got the vast majority of radio and jukebox play time, compared to songs on the B-side.

Safety performance is the A-side, it gets most of the play time during safety discussions, as it should, because of its potential negative impacts. Compliance, however, on the B-side, can also have a negative impact from citations and fines, as well as injuries, among other hidden costs. It deserves to get more play time, especially when seeking excellence in safety, which includes all four Strategic Areas of Safety.

The compliance strategic area focuses on aspects of safety involved with meeting compliance requirements, and the measures of their success. Analyzing this area helps answer the questions: "How compliant are we, how compliant do we want to be, and how can we be more compliant?"

Safety Management Systems

Having to comply with government safety regulations is a fact of life for organizations. However, the tendency for many safety professionals is to focus solely on the individual regulations. Gaps in safety compliance, unfortunately, often reveal broader issues with the systems used to manage compliance.

Dan Petersen was one of the best known and most innovative thought leaders in the safety profession. In his book, *Safety Management, A Human Approach, Third Edition*, in the chapter titled "Barriers to Safety Excellence," Petersen describes one of those barriers — government. He explains how government safety regulations have resulted in organizations developing what he refers to as "islands of safety," stand-alone programs written to comply with those regulations.[1]

A Safety Management System (SMS) is a formal, top-down, organization-wide approach to managing safety risk and regulatory compliance. It includes systematic procedures, practices, and policies for the management of safety. Multiple vendors offer computerized SMSs, which are commonly used in safety-sensitive industries. Chances are, your comany has one, or you have used one in a former safety position.

Computerized SMSs include various modules that help to simlify, integrate and automate safety. Typical modules include incident management and reporting, risk management, task management, change management, audits and inspections, and safety observations or hazard reporting. Other modules help to perform critical safety magagement processes, such as Job Hazard Analyses (JHA), Safe Work Permits (SWP), Hot Work Permits (HWP), Lockout/Tagout (LOTO), etc.

As an example, companies in industries where jobs require compliance with multiple regulations often use a Safe Work Permit (SWP) or Work Authorization process to manage job safety. This process ensures compliance with all associated safety programs and procedures. An example job is performing welding inside a confined space. Multiple safety programs apply to that work, including confined space permitting, lockout/tagout, welding safety and Hot Work Permits, respiratory protection, and personal protective equipment.

Petersen describes the safety professional's true role in relation to those "islands of safety" and management systems: "The real safety professional's role is not to write manuals. The role consists of assessing the management system, suggesting to the line organization what needs to be done, and constantly monitoring the system to ensure that it is working."[2]

When assessing compliance, safety professionals need to assess both individual safety programs and the broader safety management system(s). While an instance of non-compliance is directly associated with a regulation, it is also likely indirectly associated with a management system designed to ensure compliance. Both represent compliance gaps.

Chapter Three introduced the following three assessments for safety compliance:

1. *Regulatory Profile:* The first and foundational assessment is a Regulatory Profile or Regulatory Register. A Regulatory Profile is a prospective assessment. It looks at areas where you could receive citations and fines in the future, and the risk of receiving those.

2. *Compliance Task Profile:* The second assessment stems from the Regulatory Profile, which identifies the permits, regulations, standards, and policies that have compliance task requirements. The task profile goes one step further and lists those requirements in terms of specific compliance tasks required.

3. *Attribute Analysis:* The third assessment is an Attribute Analysis. Attribute Analysis is a retrospective assessment in the sense that it looks back on non-compliance in the past, and the reasons you experienced it.

Before describing these assessments, I must mention one of the essential elements of an effective safety management system, *auditing*, and how these assessments closely relate to it.

In his book, Peterson wrote the following about safety audits: "There are, however, many ways to build and perform audits. Some seem to make much more sense than others... In effect, the research and theory say that your audit program must look at the things that are right for your organization."[3] Generic audit checklists can contain many items not applicable to your facilities and operations while omitting others that do. Audit checklists customized to fit your company are far more effective.

There are three primary types of safety audits, as described below.

1. *Compliance Audit:* Examines how well your written programs, procedures and processes comply with laws and regulations.

2. *Program Audit:* Examines how well your organization complies with your written programs, procedures and processes.

3. *Management System Audit:* Examines the effectiveness of safety management systems and their use.

A Regulatory Profile identifies which laws and regulations you are subject to and should audit yourself against. This helps to create a custom audit checklist. It also provides a way to examine the state of your safety programs, which are written to ensure compliance. Findings from all three audit types provide excellent data for performing compliance assessments. This chapter provides three compliance assessments and presents a practical approach for performing them. The primary objective here is to help you see the value in performing them as part of your strategic planning process.

Regulatory Profile

What is a Regulatory Profile? It is a document or spreadsheet that identifies all the safety-related permits, government regulations, industry standards, and corporate policies that apply to your facility and its operations. This represents an applicability survey.

Your SMS might have a module that serves this purpose, but if not, it is critical to know what your compliance requirements are. Therefore, you will want to generate your own.

Aside from identifying regulations, standards, and policies that apply, the profile should also state the key elements required to comply with each. This represents a compliance assessment. It also serves as an overall assessment of the company's compliance.

Combining these aspects of a Regulatory Profile makes developing one a three-part process, with a fourth step added to develop any strategies based on information gathered in the first three steps.

The Regulatory Profile should also drive other aspects of your safety management system. It identifies what written programs you need, which regulations require compliance tasks, and which regulations require initial and/or periodic training. This makes your Regulatory Profile a foundational document that drives all of your compliance efforts.

As with a Risk Profile, a Regulatory Profile is something you should have regardless of the strategic planning process. The extra benefit ensures that the compliance assessment aspect identifies risks that could cause future citations and fines, loss of permits or certifications, or incidents. Compliance risks warrant consideration in your safety strategy along with performance risks. Since the Regulatory Profile is dynamic, it should be reviewed and updated at least annually, and ideally be a regular part of your strategic planning process.

If your company is certified in or pursuing certification in one of the International Organization for Standardization's (ISO) management system standards, you probably know a Regulatory Profile by another name, as described below.

1. *ISO 45001* - "Occupational health and safety management systems—Requirements with guidance for use," Clause 6.1.3, "Determination of Legal and Other Requirements," requires companies to identify all of their safety and health legal and other requirements

2. *ISO 14001* - "Environmental management systems—Requirements with guidance for use," Clause 6.1.3, "Compliance obligations," requires companies to identify all of their environmental regulatory and other obligations, along with the risks and opportunities associated with them.

It is also helpful to include a scoring system in your Risk Profile. This allows you to prioritize action plans and generate measurable targets that will imdicate when you have achieved your goals and objectives.

Part 1 - Applicability Survey

The first part of a Regulatory Profile identifies which regulatins apply to you. There are four main areas to consider, as described below.

1. *Permits:* The first area involves any operating or other permits you have, or need. Environmental permits might include air emissions, wastewater, stormwater, or hazardous waste-related permits. Some permits are both environmental and safety related, such as permits for abrasive blasting or spray-painting booths. Transportation-related permits might include wastewater, hazardous materials, or used oil. Permit non-compliance can have a negative impact on safety and the environment. If they are severe enough, the regulatory agency can shut down your operations.

2. *Regulations:* The next area to consider are the laws and regulations that you are subject to. This includes regulations, statutes, and codes at the federal, state, and local levels. You must consider regulations imposed by various government agencies, such as environmental, health, safety, fire, transportation, and labor (Workers' Compensation) agencies. Failure to comply with these regulations can cause citations or fines, and any incidents the regulations are designed to prevent.

3. *Standards:* A third area to consider is applicable industry and international standards, either those mandated by regulations, or those you voluntarily adopt for

certification. Failure to comply could cause citations or fines, loss of certification, and any incidents the standards are designed to prevent.

4. *Policies:* The last area of concern is any policies, programs, or procedures required by your organization at the regional, divisional, or corporate level, or by your customers. Failure to comply with these could have negative consequences, and should be considered in your safety strategy.

Attempting to maintain compliance with the vast number of requirements is a daunting task. This is especially true for EH&S professionals who are a department of one. Its more challenging when you have responsibility for both environmental and safety compliance. When I was in a position such as this, having a Regulatory Profile was the best way to get my arms around all that was required of me and my company.

Table 15 provides an excerpt from a compliance spreadsheet I created to serve as an applicability survey and compliance assessment. The example data is based on a facility having an anhydrous ammonia process, and includes only the survey portion for that process, omitting related federal EPA regulations.

Table 15.   Compliance Survey

			.	
Federal	Regulation	OSHA	29 CFR § 1910.119 - Process safety management of highly hazardous chemicals.	Yes
Federal	Regulation	OSHA	29 CFR § 1910.120 - Hazardous Waste Operations and Emergency Response.	Yes
Federal	Regulation	OSHA	29 CFR § 1910.133 - General [PPE] requirements.	Yes
Federal	Regulation	OSHA	29 CFR § 1910.134 - Respiratory protection.	Yes
Federal	Regulation	OSHA	29 CFR § 1910.1200 - Hazard Communication.	Yes
State	Regulation	NDEP	NRS 459.380 through 459.3874 - Chemical Accident Prevention Program	Yes
State	Regulation	NDEP	NAC 459.952 through 459.95528 - Regulation of Highly Hazardous Substances and Explosives	Yes
County	Regulation	CCFD	Chapter 13.04 - The Fire Code of Clark County - Various Sections	Yes
County	Standard	CCFD	2018 International Fire Code - Various Sections	Yes
Industry	Standard	ANSI/CGA	CGA G-2.1-2021 – Requirements for the Storage and Handling of Anhydrous Ammonia	Yes

Part 2 - Compliance Assessments

Throughout my career, I have written countless safety programs, both site-specific and standardized programs used across multiple facilities. I also performed a multitude of compliance and program audits. Through all of that, I found it very helpful to incorporate implementation guides into each program to help safety staff comply with regulations.

These guides included key areas of requirements that helped to pinpoint and track areas of non-compliance. The key areas also provide a useful way to generate a percentage-compliant metric for strategic planning. These key areas are used here to show one way to assess your level of compliance.

1. *Operating and Other Permits:* Some regulations require permits issued by a government agency, such as environmental permits. That is what this area addresses, not internal permits such as Hot Work Permits.

2. *Assessment, Survey, Monitoring:* This includes regulatory requirements to evaluate specific hazards presented by operations at a facility. Examples include PPE assessments, sound level surveys, respiratory hazard evaluations, confined space evaluations, process hazard analysis, etc.

3. *Written Program or Records:* This area addresses regulatory requirements for written programs, procedures, or processes, and/or maintain compliance records. It includes most aspects of compliance not otherwise captured by other areas listed.

4. *Initial and Periodic Training:* This area includes regulatory or company-required training. It includes training related to orientation, awareness, initial qualifications, as well as maintaining qualifications.

5. *Periodic or Episodic Tasks:* This area includes any required compliance activities, other than training, such as reporting, testing, audits, inspections, postings, etc.

6. *Medical Exams or Monitoring:* This area includes any initial, recurring, or episodic medical exams, treatment, or monitoring required by regulations. This includes respiratory exams, audiometric testing, Hepatitis "B" vaccines, post-exposure monitoring, etc.

7. *Equipment, Signs, or Supplies:* This area basically covers any goods that need to be purchased to comply with a regulation. It includes anything needed to be installed, posted, or provided to employees (e.g. PPE) for compliance with the regulations and company programs.

8. *Facility & Process Designs:* This area includes any physical features associated with the facility and its processes. It includes issues related to life safety codes, fire protection systems, walking and working surfaces, production lines, etc.

Table 16 shows key areas from table 15 assessed for compliance gaps. It omits several line items from the source table, but provides survey totals and percent non-compliant for the entire source table.

Table 16.   Compliance Assessment

	APPL	Gaps	APPL	Gaps	APPL	Gaps	APPL	Gaps
(Remainder of Table Omitted))								
29 CFR § 1910.111 - Storage and handling of anhydrous ammonia.	No	N/A	No	N/A	No	N/A	No	N/A
29 CFR § 1910.119 - Process safety management of highly hazardous chemicals.	No	N/A	Yes	Yes	Yes	No	Yes	Yes
29 CFR § 1910.133 - General [PPE] requirements.	No	N/A	Yes	Yes	No	N/A	Yes	No
29 CFR § 1910.134 - Respiratory protection.	No	N/A	Yes	No	Yes	No	Yes	No
29 CFR § 1910.1200 - Hazard Communication.	No	N/A	Yes	Yes	Yes	No	Yes	Yes
	5		47		75		90	
Percent Non-Compliant		0.0%		17.0%		12%		8.9%

	APPL	Gaps	APPL	Gaps	APPL	Gaps	APPL	Gaps
29 CFR § 1910.111 - Storage and handling of anhydrous ammonia.	No	N/A	No	N/A	Yes	No	Yes	No
29 CFR § 1910.119 - Process safety management of highly hazardous chemicals.	Yes	No	No	N/A	Yes	No	Yes	No
29 CFR § 1910.120 - Hazardous Waste Operations and Emergency Response.	Yes	No	Yes	No	Yes	No	No	N/A
29 CFR § 1910.133 - General [PPE] requirements.	No	N/A	No	N/A	Yes	No	Yes	No
29 CFR § 1910.134 - Respiratory protection.	Yes	No	Yes	No	Yes	No	Yes	No
29 CFR § 1910.1200 - Hazard Communication.	No	N/A	No	N/A	Yes	No	Yes	No
(Remainder of Table Omitted)								
	90		20		91		36	
Percent Non-Compliant		16.7%		10.0%		17.6%		5.6%

Information for performing the assessment can come from a variety of sources, including first-hand knowledge, incident investigations, written program reviews, findings from agency audits and inspections, findings from internal audits and inspections, and even safety suggestions and observations. The objective is to identify all gaps in compliance.

In table 16, you can see that some of the key areas have a higher percent non-compliant. For strategic planning, those areas warrant a deeper dive to determine what the underlying causes are.

Table 17 provides the overall percentage noncompliant or compliant. These totals can represent the overall compliance status, a useful metric for performance measures.

Table 17.   Compliance Assessment Score

Permit, Regulation, Standard, Policy	Compliance Status		
	# Req'd	# Gaps	% Gaps
(Remainder of Table Omitted)			
29 CFR § 1910.111 - Storage and handling of anhydrous ammonia.	2	0	0%
29 CFR § 1910.119 - Process safety management of highly hazardous chemicals.	6	2	33%
29 CFR § 1910.120 - Hazardous Waste Operations and Emergency Response.	5	0	0%
29 CFR § 1910.133 - General [PPE] requirements.	3	1	33%
29 CFR § 1910.134 - Respiratory protection.	7	0	0%
29 CFR § 1910.1200 - Hazard Communication.	5	2	40%
(Remainder of Table Omitted)			
Total	454		

Percent Non-Compliant   **13%**

Part 3 - Corrective Actions

Having identified all of your compliance obligations, and assessed them for gaps, it is time to determine what corrective actions are needed. First, it helps to document how you discovered the gap, the source of information, for future reference. It is also helpful to include the risk to the organization, which helps to determine which gaps warrant inclusion in your safety strategy. Including a severity score would also work.

Table 18 describes a few of the corrective actions for select regulatory gaps in the anhydrous ammonia example. The gaps individually warrant corrective actions. Collectively, however, they might suggest a broader concern related to the "islands of safety" that Peterson mentioned.

Table 18.   Corrective Actions

29 CFR § 1910.119	1. The Management of Change (MOC) process failed to recognize incompatible seal material in a new safety valve that was not a replacement in kind. 2. Operators were not trained on the system changes prior to system startup.	Incident Investigation	Inadvertent release of anhydrous ammonia resulting in death or serious physical harm.	Revise the MOC process to include the detailed review and approval of new components for compatible materials, and a sign-off for pre-startup training being completed.
29 CFR § 1910.133	The PPE Assessment failed to include the Level A totally-encapsulating chemical protective suit required for emergency response operations.	Compliance Audit	Operators failing to don the correct PPE resulting in death or serious physical harm.	Revise the PPE Assessment to include Level A chemical suits and review for other PPE missed in the assessment.

The OSHA Process Safety Management (PSM) regulation applies to extremely hazardous substances (EHS) as listed by the EPA, and non-compliance can cause a dangerous release of those substances. While the PSM regulation addresses many of the safety requirements associated with the covered process, it is not the only regulation that applies to it. Therefore, compliance cannot be judged by just one regulation, regardless of how all-encompassing it is.

The comprehensive Regulatory Profile brought the applicable regulations into one assessment. The existing safety management system allowed compliance gaps associated with process safety. What may have been missed during a PSM program audit, or the required Process Hazard Analysis (PHA), the Regulatory Profile brought to light.

Part 4 - Strategy

A Regulatory Profile provides a few sources of strategies and strategic plans. One source comes from the higher-level look at overall compliance associated with the covered process. Another source comes from the gaps identified under individual regulations.

Referring back to table 16, one example is finding multiple gaps in the equipment, signs, and supplies area. Of all the applicable permits, regulations, standards, and policies, 17.6% had compliance gaps in this area. Suppose internal inspections identified a general lack of safety supplies. Diving deeper into the issue, you might discover that there is no safety supply inventory process, or an inadequate one.

The assessment, survey, and monitoring area had the second highest non-compliance percentage with 17% of requirements having gaps. This could be because your written safety programs did not include those requirements, there is a lack of knowledge among your EH&S staff, or both. Analyzing competency should be part of your strategy to improve in that area of compliance. (Chapter Nine will address analyzing competency.)

The third-highest non-compliance percentage was in the periodic and episodic tasks area, with 16.7% of requirements having gaps. Developing the Compliance Task Profile described below should be part of your strategy.

Another source of strategies is the overall Risk Profile. What is the general state of your compliance efforts? Does the overall percentage-compliant score of 87% (100%-13%) seem low to you? If so, is there an underlying reason for poor compliance? It might be a culture issue where people perceive compliance being less important, or the organization does not provide adequate resources for safety. Analyzing your safety culture should be part of your strategy. (Chapter Ten will address analyzing culture.)

The Risk Profile brings issues like these to light. As the old saying goes, "Sunlight is the best disinfectant." It is up to you to determine the underlying cause and address it.

Compliance Task Profile

The Compliance Task Profile picks up where the Regulatory Profile left off regarding compliance tasks. The Risk Profile identifies which permits, regulations, standards, or policies are applicable and have required tasks associated with them, but it does not list those tasks. A Compliance Task Profile lists what your organization's specific task requirements are.

Having a comprehensive list of required compliance tasks allows you to assess how well you are doing at performing all of those tasks. This is the primary value of having a Compliance Task Profile. What percentage of tasks are you completing? By not performing tasks, you could be subjecting yourself to future citations and fines, or worse, serious injuries.

A Compliance Task Profile also provides the foundation for setting up the system you use to manage those tasks. This could be the task management component of Safety Management System (SMS) software, or similar. It could also be a compliance calendar. To ensure safety compliance excellence, you must have a system in place for managing compliance tasks.

Part 1 - Profile Development

The best way to develop your profile is to go directly to the source. There are various types of tasks to look for, including: reporting, testing, audits, inspections, postings, etc. Look for both periodic and episodic tasks. Episodic tasks are required following an event, such as notifying OSHA of reportable injuries or reporting environmental releases. Episodic tasks may also include equipment pre- and/or post-use inspections.

When developing your profile, review all the potential sources that were described above.

1. *Permits:* While some permits only have a renewal date to be concerned with, others have multiple conditions to comply with. This includes site-specific permits, general permits, and permits by rule. Site-specific permits will provide the required tasks for your site. If you are covered by a general permit, such as a Multi-sector General Permit (MSGP) for stormwater, the permit includes required compliance tasks. If you fall under a permit by rule, the rule will state required compliance tasks.

2. *Regulations:* OSHA, EPA, DOT, and other regulations often have required tasks specifically stated in the regulation. Sometimes, however, the text implies a task, like first aid kit inspections to ensure adequate supplies are readily available. Sometimes an industry standard is incorporated by reference and you must read the industry standard to get the details. You must also consider state and local statutes and codes that you are subject to, which may also have required tasks.

3. *Standards:* Industry and international standards do an excellent job stating their required compliance tasks. Keep in mind that OSHA often incorporates industry standards by reference. OSHA lists all industry standards incorporated by reference (IBR) in 29 CFR 1910.6. Sometimes, OSHA allows for voluntary compliance with industry standards to meet the requirements of its regulations. Exit-route regulations are a prime example. You can comply with the NFPA 101, Life Safety Code, 2009, or the exit-route provisions of the International Fire Code, 2009, instead of the OSHA regulations, as stated in 29 CFR 1910.35.

4. *Policies:* Your organization's policies, programs, or procedures may include tasks not found in the sources described above. They could be related to internal reporting, or represent best management practices not included in or required by other sources. They might also include differences from regulatory requirements, such as performing a task more frequently.

After you develop your initial Compliance Task Profile, you will need to watch for changes to permits and regulations and update your profile accordingly. Regarding industry standards, those incorporated by reference in OSHA regulations have more current revisions, but technically you only need to comply with the specific revision incorporated by reference (e.g. NFPA 101, 2009 revision.)

Part 2 - Compliance Assessments

After developing your Compliance Task Profile, the first assessment you should perform is ensuring that your management system includes each requirement. If your organization already uses Safety Management System (SMS) software, or something similar, with a task management component, ensure that all of your required periodic tasks are included. Do the same for whatever form of compliance calendar you may use instead. Also, ensure that your written safety programs include their respective requirements.

The second assessment you should perform is comparing what you are required to do against what you are actually doing. If your SMS or similar has a task management component, it should have records of task completion that you can analyze. Like other assessments, it is helpful to include a scoring system to give you a measure of performance. In this case, it should measure the percentage of compliance tasks completed.

You should calculate a completion rate metric for individual tasks to identify specific problem areas. However, an overall completion rate will serve as a good strategic planning performance measure. In my experience, target completion rates of 90% or even 95% are achievable and make for inspirational targets. If your overall completion rate is lower than the desired rate, improvement plans should be included in your strategic safety plan.

One of the primary aspects of safety compliance is performing required tasks. Failing to perform those tasks can cause injuries, or result in citations and fines. To avoid those, it is important to first know what the requirements are, and then to have a management system that ensures you are completing them.

Attribute Analysis

An Attribute Analysis is another excellent analytical tool for uncovering compliance gaps. It was presented in the previous chapter. Here we will discuss some of the unique aspects of a Compliance Attribute Analysis

Attributes & Attribute Selections

Similar to other data, you need a distinct set of attributes to analyze compliance. Table 19 provides an example of eight attributes, and their list of selections, that you can use to categorize and analyze compliance failures.

Table 19.   Compliance Attribute Lists

Environmental	Permit	Operating or Other Permits	Citation
Health & Safety	Regulation	Assessment, Survey, Monitoring	Agency Audit or Inspection
Transportation	Ind. Standard	Written Program or Records	Internal Audit or Inspection
Labor	Corp. Policy	Initial or Periodic Training	Program Review
		Periodic or Episodic Tasks	Suggestion or Observation
		Medical Exams or Monitoring	Incident Investigation
		Equipment, Signs, or Supplies	
		Facility & Process Design	
CAT 1 (Major)	Lack of Resources	N/A (Administrative)	EH&S Staff
CAT 2 (Minor)	Lack of Knowledge	Office Building	Management
Observation	Lack of Supervision	Main Shop	Administration
Opportunity for Improvement	Not Communicated	Machine Shop	Shop Supervisor
Compliant	Misinterpretation	Yards & Grounds	Field Supervisor
Noteworthy Effort	Time Pressure	Multiple Locations	Technician
	Poor Housekeeping	Field Service Project	Other Position
	Unintended Oversight		
	Intentional Violation		

Notice that the compliance areas list is identical to the key areas of compliance used in the Regulatory Profile above. This provides a comparison between the two sets of information that might prove useful. For example, if your highest percentage of non-compliance is in the equipment, signs, and supplies area, it helps to know that the highest number of regulations applicable to you are also in that area.

The finding category provides a severity score, of sorts, for compliance audits. An ISO certification body uses the list shown here that includes both negative and positive categories. The bottom three are positive finding categories where the company complied with the requirement but could still improve their management systems or efforts, or showed best-in-class performance. The observation category are areas that were close to becoming a non-compliant finding, or, given additional scrutiny, could be non-compliant. I recommend including both negative and positive categories. Audit results are better received when they reflect overall compliance, which requires stating the good, the bad, and the ugly.

Attribute Data

As seen in the attribute information source in table 19, there are a variety of sources that you can get data from for your compliance Attribute Analysis. The data for these sources may be in document format, either electronic or hard copy, or in spreadsheet or database format, such as a Corrective and Preventive Action (CAPA) Plan.

If your company uses auditing software, this should provide you with a good primary source of data, whether it is in spreadsheet or database format. If you need to augment this with additional attributes, you will need to download the raw data from that application into spreadsheet format and add other attributes to it, providing a secondary source of data.

If your company has certification in one of the International Organization for Standardization's (ISO) management system standards, you likely already have one source of data in what is commonly called a Corrective & Preventive Action (CAPA) Plan, though later versions of the standards omit references to preventive actions. A CAPA Plan complies with the ISO requirements described below.

1. *ISO 45001* - "Occupational health and safety management systems—Requirements with guidance for use," Clause 6.2.2, "Planning to achieve OH&S objectives," requires companies to maintain documented information on their objectives and plans to achieve them.

2. *ISO 14001* - "Environmental management systems—Requirements with guidance for use," Clause 6.2.2, "Planning Actions to Achieve Environmental Objectives," requires companies to provide an action plan describing how their environmental objectives are achieved.

As with data from your auditing software, your ISO CAPA Plan may need to be augmented with additional attributes to make your analysis more effective. A secondary data spreadsheet will serve this purpose.

As with the levels of warfare described in Chapter Two, safety strategies include strategic, operational, and tactical plans. ISO management systems typically generate actions related to operational or tactical plans. For strategic planning, you are also concerned with strategic plans. During an Attribute Analysis, you will generate insight needed for all three levels of plans.

Keep in mind that compliance data has the same concerns regarding quality of data that performance data does, including: *incompleteness, inconsistency*, and *noise*. Compliance auditing data is especially prone to inconsistent data when there are several auditors involved. The best way to minimize inconsistencies is to have a written audit protocol that clearly defines the attributes and their list of selections. This allows all auditors to sing from the same sheet of music.

Data Mining

Mining compliance data is the same as performance data, especially if you use the same or a similar database or spreadsheet. Table 20 comes from a secondary data spreadsheet I created to perform Attribute Analysis with, similar to the one used for performance. The table looks at the five most valuable attributes for performing a compliance Attribute Analysis, in my opinion. They represent hypothetical findings from internal regulatory compliance audits and inspections for a regional group of business units with similar operations. The data set contains one hundred records. (NOTE: Since the total count is one hundred, individual counts serve as percentages as well.)

Table 20.   Top 5 Compliance Findings

Corporate EH&S Standard	11	Equipment, Signs, or Supplies	40	Administrative (Location N/A)	50	EH&S Staff	76	Lack of Knowledge or Skills	55
29 CFR § 1910.1200 - Hazard Communication	8	Written Program or Records	26	Main Shop	43	Shop Supervision	19	Unintended Oversight	18
29 CFR § 1910.37 - Maintenance, safeguards, and operational features for exit routes	7	Periodic or Episodic Tasks	12	Multiple Locations	3	Management	2	Not Stated or Communicated	10
29 CFR § 1910.212 - General requirements for all machines.	7	Assessment, Survey, Monitoring	8	Office Building	2	Administrative Personnel	1	Poor Housekeeping Practices	7
29 CFR § 1910.147 - The control of hazardous energy (LOTO)	5	Facility & Process Design	7	Yard & Grounds	1	Other Positions	1	Intentional or Known Violation	7

The table presents data from a single-level analysis of each attribute with no cascading of attributes. Column A, Regulation, provides the top five regulations along with a corporate standard that were the subject of the audits and inspections. Non-compliance with corporate standards had the most findings, as opposed to specific regulations.

Forty percent of findings were in the equipment, signs, and supplies compliance area, which likely reflects the fact that it has the highest number of requirements, as suggested above. It stands to reason it would have a comparably higher percentage of findings. Twenty-six percent of findings related to written programs or records, which may or may not provide useful strategic insight based on further examination of the individual findings.

Fifty percent of the findings were administrative, stemming from program and records reviews, as opposed to physical or conditional findings discovered during facility inspections. Those would include findings in the compliance areas related to initial assessments or surveys, written programs, compliance tasks, and compliance training. The first level of analysis might not provide actionable insight. Table 21 provides the results of a cascading analysis, as described previously, starting with the EH&S Staff. (NOTE: The percentages are percent of all findings.)

Table 21.   EH&S Staff Findings

Organizational Position	%	Causal Factor	#/%	Compliance Area	%
EH&S Staff	76	Lack of Knowledge	47	Written Program or Records	13
		Lack of Oversight	15	Equipment, Signs, or Supplies	10
		Not Stated or Communicated	6	Periodic or Episodic Tasks	8
		Intentional Violation	5	Assessment, Survey, Monitoring	6
		Misinterpretation	3	Initial or Periodic Training	5

A deeper dive into the data reveals that almost half of all audit findings involved a lack of knowledge in the EH&S staff. Why is that the case? You should examine this further to help identify strategies for improvement.

A final note involves past regulatory citations. While those should be included in your Attribute Analysis, some key citations might get overlooked. Those are citations that were issued within your company, but involved other locations. They respresent a risk to your location in the form of a costly repeat citation.

When OSHA issues a citation, it is issued to the company. Any subsequent violations where the hazard type is substantially the same, can be cited as a repeat vilation. The violation does not need to occur at the same facility, it just needs to occur within the same company. For that reason, citations anywhere within the company should be included.

I trust these examples give you a sense of how valuable Regulatory Profiles and an Attribute Analysis can be in this strategic area. It brings to light the aspects of safety compliance where gaps exist, which may warrant inclusion in your safety strategy, beyond individual corrective actions.

The benefit of a Regulatory Profile, Compliance Task Profile, and Attribute Analysis is that they present data in different ways. That includes data related to what non-compliance could occur in the future, and instances of non-compliance that have occurred in the past, based on audit findings and other sources.

The next chapter will examine the analysis of competence, the third strategic area of safety. This analysis will include another method that picks up where the Regulatory Profile left off, and it will introduce a new form of assessment that is used for both the compliance and culture strategic areas.

Endnotes

1. Dan Petersen, Safety Management, a Human Approach, Third Edition, (Des Plaines: American Society of Safety Engineers, 2001), 327.

2. Ibid.

3. Ibid., 27-28.

Chapter Eight:
Competence Analysis

Competency absolutely is required as part of a sustainable safety culture. — Carl Potter

Are you aware of your areas of competency? What areas are you competent in, and what areas are you not? How about the people you work with? There are four stages of competency; the first two involve incompetence, whereas the next two involve competence. Knowing someone's areas of competence or incompetence is important in safety.

It is the first two stages we are concerned with here, especially the first one. The first two stages are:

1. *Stage 1. Unconscious Incompetence:* People who don't know what they don't know are in this stage. These people can have a detrimental impact on your safety program because they don't know when to get help in a given area of incompetence, and could make unsafe decisions on their own.

2. *Stage 2. Conscious Incompetence:* People who know what they don't know are in this stage. These people at least know when to get help in a given area of incompetence, and are less likely to make unsafe decisions.

Competence analysis is designed to help those in the first two stages. It is designed to help get them to the third stage, conscious competence, and eventually to the fourth stage, unconscious competence (mastery).

The competence strategic area focuses on aspects of safety involved with developing a higher level of safety competency throughout the organization. Analyzing this area helps answer the questions, "How competent are our people, how competent do we want them to be, and how can we improve their competency?"

Many of the Occupational Safety and Health Administration (OSHA) regulations require employers to provide employees with training to ensure that they are aware of, and knowledgeable in, the subject addressed in the regulations. OSHA also requires employers to have a "competent person" to oversee some safety programs. OSHA defines a "competent person" in 29 CFR 1926.32(f):

> Competent person means one who is capable of identifying existing and predictable hazards in the surroundings or working conditions which are unsanitary, hazardous, or dangerous to employees, and who has authorization to take prompt corrective measures to eliminate them.

The need for safety knowledge and competence within the organization, however, goes well beyond specific OSHA requirements for training or having a "competent person." Everyone in the organization, from the boardroom down to the shop floor, needs to have the safety competencies commensurate with their safety roles and responsibilities.

Carl Potter is an internationally recognized safety expert at The Safety Institute. In an *EHS Today* article titled "Are You Competent When It Comes to Safety," Porter asks the following question: "Have you ever wondered what it would be like to have a fully competent workforce, where every person was fully capable of performing work safely?"[1] While Potter recognizes that as being a Utopian concept, he noted it is worth pursuing. He later states: "Competency absolutely is required as part of a sustainable safety culture."[2]

If you want to achieve safety excellence, then you must address safety competence.

Chapter Three introduced the two assessments below for the competence strategic area that will help you identify the safety competencies required for your organization, and who may be lacking in these.

1. *Safety Training Profile:* The first assessment examines the safety training requirements for all applicable positions within the organization. It identifies the training mandated by regulations, as well as other training deemed necessary

by the organization. It then assesses the status of that training to identify areas where improvements are needed to ensure the highest level of safety awareness and qualification.

2. *Competency Assessment:* The second assessment evaluates the competency of safety professionals and other key positions in the organization. It identifies areas where improvements are needed to achieve an appropriate level of proficiency in safety within those roles.

These assessments offer an excellent source of insight into what objectives you should include in your safety strategy to achieve excellence in safety competence. This chapter provides an introduction to these assessments and presents a practical approach for performing them. Like other areas, the primary objective is to help you see the value in performing them as part of your strategic planning process.

Safety Training Profile

A Safety Training Profile is like the Compliance Task Profile described in the previous chapter. The profile picks up where the Regulatory Profile left off, identifying what your organization's specific safety training requirements are. The Compliance Assessment portion of the Regulatory Profile identified which applicable permits, regulations, standards, and policies have training requirements, but did not list specific training.

Your organization may have a Safety Training Profile if you are using a Safety Management System (SMS) that has a training management component, or a standalone Learning Management System (LMS).

A Safety Training Profile describes all of your required training courses. Table 22 provides an excerpt from the first half of a spreadsheet I used to help business units without SMS or LMS software develop a training matrix to manage their safety training.

Table 22.   Safety Training List

EHS Category & Topic	APPL	Regulation	Ret. Freq	Training Type	Training Hours
HS - Safe Work Permitting					
Risk Assessment and JSA	Yes	Policy	N/A	Classroom	1.00
Work Authorization and Permitting	Yes	Policy	N/A	Classroom	1.00
Hot Work Permits & Fire Watch	Yes	29 CFR 1910.119	1 YR	Classroom	1.00
Lockout/Tagout - Awareness	Yes	29 CFR 1910.147	1 YR	CBT	0.50

Lockout/Tagout - Authorized Employee	Yes	29 CFR 1910.147	1 YR	CBT	1.00
Lockout/Tagout - Supervisor	Yes	29 CFR 1910.147	1 YR	Classroom	2.00
Confined Space Entry - Entrant	Yes	29 CFR 1910.146	1 YR	CBT	1.00
Confined Space Entry - Attendant	Yes	29 CFR 1910.146	1 YR	CBT	2.00
Confined Space Entry - Entry Supervisor	Yes	29 CFR 1910.146	1 YR	Classroom	4.00
Atmospheric Testing	Yes	29 CFR 1910.146	1 YR	Classroom	1.00
Electrical Safety - General	Yes	29 CFR 1910.332	Initial	CBT	0.50
Electrical Safety - Unqualified Person	Yes	29 CFR 1910.332	Initial	Classroom	2.00
Electrical Safety - Qualified Person	Yes	29 CFR 1910.332	Initial	Seminar	8.00

The categories in this list included corporate policies, environmental regulations, health and safety regulations, and industry or international standards. Safe Work Permitting, a subcategory under the health and safety category, is used for demonstration purposes. The complete list included over one hundred training topics, including international requirements.

Training types included classroom, computer-based training (CBT) and seminars, as seen in the table, as well as safety meetings and toolbox talks for less involved topics. This helped business units to map out their classroom training, safety meeting, and toolbox talk topics for the year. Table 23 provides an excerpt from the second half of the spreadsheet, identifying training required for each position.

Table 23.   Safety Training Assignments

HS - Safe Work Permitting						
Risk Assessment and JSA	Yes	Yes	Yes	Yes	Yes	Yes
Work Authorization and Permitting	Yes	Yes		Yes	Yes	
Hot Work Permits & Fire Watch	Yes	Yes	Yes	Yes	Yes	Yes
Lockout/Tagout - Awareness			Yes			
Lockout/Tagout - Authorized Employee	Yes	Yes		Yes	Yes	Yes
Lockout/Tagout - Supervisor	Yes	Yes		Yes	Yes	
Confined Space Entry - Entrant	Yes	Yes	Yes	Yes	Yes	Yes
Confined Space Entry - Attendant	Yes	Yes	Yes	Yes	Yes	
Confined Space Entry - Entry Supervisor	Yes	Yes				
Atmospheric Testing	Yes	Yes		Yes	Yes	
Electrical Safety - General			Yes			
Electrical Safety - Unqualified Person	Yes	Yes		Yes		Yes
Electrical Safety - Qualified Person					Yes	

Whatever your organization uses to track training completion, whether it is an SMS, LMS, or other form of tracking, it should track how well training is being completed. It should provide a leading indicator metric, percent completion or percent on-time completion. Why the latter? Annual traing required by OSHA is due within 365 days.

Regarding training compliance rates, my experience has been that target completion rates of 90% or even 95% are achievable and make for inspirational targets. This metric, is valuable to a strategic safety plan for measuring the overall level of training. That metric, in a sense, indicates how well the organization is doing in developing and managing a good portion of the functional competencies related to safety, as described below.

Competency Assessment

In keeping with Voltaire's recommendation about defining terms, before discussing Competency Assessments, the terminology involved should be clearly defined. Above we looked at the OSHA definition of a "competent person." The word *competent* is an adjective that describes someone having the knowledge, skills, and ability to do a job well; well enough to meet basic standards. By comparison, someone who is proficient can do the job *very* well; better than a competent person, due to having more experience.

Competence is a noun that refers to the ability of a person to do the job well. A competent person, by definition, has competence in the specific job they can do well. *Competency* is a similar noun that has a subtle difference in meaning. The term is usually used to refer to a professional skill or knowledge, or a specific set of skills or knowledge, that is required for a particular role. Being well-versed in office suite software, for example, is a competency. This is how the word is used in terms of a Competency Assessment. A Competency Assessment, therefore, is an evaluation of whether a person has a specific set or sets of skills or knowledge required for their role in the organization.

An organization can assess safety competency for any position. Ideally, it would be beneficial to perform one for managers and supervisors overseeing safety-intensive operations, especially if they wear safety as a second hat. The practicability of this depends on the size and resources of the organization. You should, at the very least, assess the competency of people in full-time safety positions.

Types of Competencies

Before discussing the analysis of competency, it is helpful to briefly review the various types of competencies. Below are five basic types of competencies.

1. *Core Competencies:* These are typically described in one of two way. They are described by some as the foundational skills needed to ensure consistency of behavior in the workplace, such as integrity, empathy, active listening, teamwork, etc. These are similar to behavioral competencies, which are sometimes listed seperately. Others describe core competencies as those unique qualities or strengths that your employees possess that sets your organization apart from its competitors, such as innovation, customer-focus, adaptability, etc.

2. *Functional Competencies:* These relate to your role in the organization, which, for safety professionals, would include a variety of safety topics. Examples include everything from knowing safety regulations and their requirements, to knowing how to perform safety tasks like audits and inspections, or incident investigations.

3. *Technical Competencies:* This refers to the specialized knowledge required to use the software systems and advanced tools or processes associated with a particular role. Knowledge and skills in using the (SMS software, the LMS software, or the incident reporting system are some examples. It would also include operating specialized safety euipment such as a multi-gas detector docking station that automatically performs daily calibration checks.

4. *Leadership Competencies:* These competencies focus on the ability to lead a function, guide teams, and manage people. Specific competencies include strategic planning, team-building, and coaching or mentoring, etc..

As you can see, safety competencies can run the gambit Developing those competencies would include formal training, hands-on experience, and coaching or mentoring. This chapter, however, focuses on the assessment of safety competency within the organization.

Competency Framework

The first step in assessing competency is developing a competency framework. A competency framework is a model that describes the various competencies someone must have for a specific role. Some industry organizations have published guidance on how to develop a safety competency framework, but few actually provide one.

The Institution of Chemical Engineers (IChemE) is a United Kingdom-based qualifying body and society for chemical, biochemical and process engineers. Their IChemE Safety Centre provides a Process Safety Competency model that includes all roles involved with process safety. It is a very detailed model that clearly defines each area of competency and the specific competencies needed for each role.

The Institution of Occupational Safety and Health (IOSH) developed a model specifically for safety professionals. IOSH is the world's largest membership body for safety and health professionals, also based out of the United Kingdom. The IOSH Competency Framework is comprised of sixty-nine competencies organized under twelve areas that fall within one of three primary categories: *technical*, *core*, and *behavioral* competencies.[3] A detailed description of the competencies is available on the IOSH website. The full model, however, is only available to IOSH members.

Chapter Ten will describe a Safety Management Framework. The list below provides the twelve elements used in that framework, each of which has five sub-elements to define them further. Safety professionals need to have various levels of competency in each of these elements:

1.	Visible Leadership	7.	Safe Work Processes
2.	Management Systems	8.	Safe Work Behaviors
3.	Competent Personnel	9.	Audits & Inspections
4.	Culture & Communications	10.	Change Management
5.	Contractor Management	11.	Incident Management
6.	Risk Management	12.	Performance Analysis

Competency Scale

The next step in assessing competency is developing a competency scale. Trade crafts use a well-known scale: *apprentice*, *journeyman*, and *master*. That scale dates back to the craft guilds of the Middle Ages. An apprentice is someone new to the trade who is learning under the tutelage of a journeyman or master. A journeyman is someone who has completed apprentice training and is skilled in the trade, but not experienced or knowledgeable enough to start their own business. A master has enough knowledge and experience to start their own business. In modern times, requirements for obtaining a journeyman and master license include examinations and certain hours worked in that particular trade.

Some models, like the IChemE model, add a fourth level of competency at the beginning of the scale, the *awareness* level. The IChemE model combines mastery and expert at the end of the scale, whereas other models split those two to create a five-level scale.[4] The scale used in this example has five levels, and is based on the Dreyfus Model of Skill Acquisition:

1. *Novice::* A person with little or no training or experience in the subject who is just beginning to learn and gain experience in the subject. (e.g. Apprentice)

2. *Advanced Beginner:* A person in training and gaining initial experience, who has developed some skill and knowledge in the subject, but needs more hands-on experience to become competent. (e.g. Safety Specialist)

3. *Competent:* A person who completed training and has enough experience to be knowledgeable and skilled in the subject. (e.g. Safety Coordinator)

4. *Proficient:* A person who has received advanced training or education and has sufficient experience to have mastered the subject. (e.g., Safety Manager)

5. *Expert:* A person with advanced education, comprehensive knowledge, and extensive experience in the subject, or is recognized by a certifying authority. (e.g. Certified Industrial Hygienist)

Using the Competency Framework and scale, the organization must then decide which level an individual in each job position should achieve. This might differ between the elements. For example, a Safety Coordinator might need to be competent (Level 3) in the management systems element, because of writing and auditing safety programs. However, they only need to be an advanced beginner (Level 2) in the visible leadership element, because of just starting to learn leadership responsibilities..

Determining the target levels of competency for the various positions should be performed by a group representing all safety positions. This helps to ensure the accuracy of the assessment tool, and promotes agreement from those assessed against it.

Assessing Competency

The Competency Assessment process involves selecting the level of competence of an individual under each element or sub-element in the framework. Assessments at the sub-element level provide results that are more meaningful since sub-elements are more closely related to specific competencies.

After completion, the process compares an individual's assessed level against the target level for their position. Where the target level has not been achieved, personal improvement plans should be developed.

You can perform Competency Assessments using one or more of three methods.

1. *Self-Assessment:* While not recommended to be used alone, Competency Assessments are a great way for safety professionals to reflect on their own professional strengths and weaknesses and create a personal professional development plan. Self-assessments provide the individual with input into the assessment, similar to personal input to performance appraisals.

2. *Manager Assessment:* Manager should perform a Competency Assessment of subordinates. The manager, presumably, has achieved higher levels of competency and knows better than the individual what it takes to achieve them.

3. *Team Assessment:* Some organizations use a team approach, like in a 360-degree feedback system, having an individual's supervisor, co-workers, and even subordinates, take part in the assessment. When using this method, scores from all the assessments are typically averaged to determine the individual's score.

Table 24 is a Competency Assessment for a Safety Coordinator, showing the first four (foundational) elements from the Safety Management Framework. It includes target scores for the various safety positions and the individual's actual scores.

Table 24.   Competency Assessment–Safety Coordinator

Element	Sub-Elements	Target Scores / Levels					Actual Scores	
		Beginner	Specialist	Coordinator	Manager	Director	Level	Average
Visible Leadership	Safety Strategy						1	
	Clear Expectations						1	
	Resources & Support	1	2	3	4	5	2	1.4
	Visible Presence in the Workplace						2	
	Safety Leaders at All Levels						1	
Management Systems	Safety Management System						3	
	Policies, Procedures & Processes						4	
	Task Management System	1	2	3	4	5	3	3.0
	Awareness Campaigns						2	
	Monthly Reporting						3	
Competent Personnel	Organizational Structure						3	
	Recruiting & Selection Practices						2	
	Training Programs	1	2	3	4	5	3	2.8
	Training Management System						4	
	Professional Mentoring						2	

Element (continued)	Sub-Elements	Target Scores					Actual Scores	
		New Hire	Specialist	Coordinator	Manager	Director	Level	Average
Culture & Comm.	Culture Improvement Projects	1	2	3	4	5	3	3.8
	Caring Conversations & Coaching						4	
	Safety Committees & Meetings						4	
	Toolbox Talks & Job Briefings						4	
	Safety Suggestion Program						4	
(Portion of Table Omitted)								
Assessment Totals		15	24	36	45	54		30

Visible leadership happened to have the element score furthest from its target score. One explanation might be the individual could not attend the organization's annual safety leadership seminar, and therefore could not start getting knowledge and experience in those areas. This illustrates why Competency Assessments should never be used as a performance evaluation tool, only a professional development tool. A lack of competence is not necessarily a performance issue.

For these elements, the individual scored thirty out of a target score of thirty-six, or 83% of the target competence level for a Safety Coordinator. If the organization set an overall target of 90% competence level, professional development plans would be warranted for the individual.

Assume the average competence level for all Safety Managers combined was only 75%, with a target of 90%. A strategy for improving the competency of Safety Managers would be needed to get them to the targeted level.

Diving deeper into the assessments might reveal a trend related to specific elements having lower scores that would provide the operational and tactical plans for that strategy. This is the benefit of Compliance Assessments, gaining insight into the competency of the assessed group of individuals, and determining which actions will improve it, if needed.

Competency Target Levels

In order to complete a Competency Assessment, you must determine the target levels of competency for each position, before each individual is assessed against it. In this example, all target scores, or the levels of expected competency, fell directly in line with the five positions. There is nothing stating that those levels couldn't be adjusted.

For example, under the visible leadership element, you may decide that a New Hire does not need to achieve level one, but a Safety Specialist does.

The tools used to define the target levels can be as simple or as complex as you deem necessary. An example of a comprehensive model can be seen in the IChemE Safety Centre's Process Safety Competency model. Appendix 1: Competency matrix, in their IChemE Safety Centre Guidance document, provides an excellent example of a matrix showing the competency level targets for 24 process safety positions in each of their eighteen competecy elements.[5]

This chapter presents a more simplified model using descriptors for elements of the Safety Management framework. This same method will be used in the next chapter, as well, for assessing the safety program and/or culture. Hence the use of it here.

To make the assessment as objective as possible, descriptors would be included for each element and position. Table 25 provides a list of descriptors as an example.

Table 25.   Competency Level Descriptors

Element	Target Scores				
	Beginner (Novice)	Specialist (Advanced Beginner)	Coordinator (Competent)	Manager (Proficient/Expert)	Director (Expert/Dept. Head)
Visible Leadership	Should be aware of the safety strategy. No other leadership expectations.	Helps develop safety strategies. Provides resources and support based on hazards.	Helps develop safety strategies. Determines controls needed for safety hazards.	Oversees strategy execution. Sets clear expectations regarding safety. Develops safety leaders. Is visibly present in the workplace.	Facilitates strategy development. Sets clear expectations regarding safety. Ensures financial resources. Is visibly present in the workplace.
Management Systems	Starts learning how to use management systems. Familiarizes self with policies, procedures, and processes.	Uses management systems. Knows policies, procedures, and processes. Provides input to monthly reports.	Maintains management systems. Helps develop procedures, and processes. Provides input to monthly reports.	Develops or acquires management systems. Helps determine safety policies. Generates monthly reports.	Approves management systems. Approves or sets safety policies. Approves monthly reports and shares with executive leadership.
Competent Personnel	Serves as a trainer's assistant to learn the training process. Improves knowledge in various safety programs.	Delivers training. Works at becoming a "Competent Person" for some programs. New Hires to develop their competency.	Develops training program and delivers training. Fulfills "Competent Person" roles for various programs. Mentors Safety Specialists.	Oversees training program. Involved with hiring or promoting safety staff. Mentors Safety Coordinators.	Establishes the organizational safety structure and staffing levels. Mentors Safety Managers.

Element	Target Scores (continued)				
	Beginner (Novice)	Specialist (Advanced Beginner)	Coordinator (Competent)	Manager (Proficient/Expert)	Director (Expert/Dept. Head)
Culture & Comm.	Participates in culture improvement programs. Assists with toolbox talks and safety suggestion program.	Participates in culture improvement programs. Coaches employees in safety. Coordinates toolbox talks and safety suggestion program. Coaches employees in safety.	Helps develop culture improvement programs. Provides job safety briefings. Coaches employees in safety. Member of the Safety Committee.	Develops culture improvement programs. Chairs the Safety Committee. Monitors safety suggestions for needed administrative improvements.	Oversees culture improvement programs. Coordinates organization-wide safety communications.

Table 25 gives you an idea of how to develop descriptors for your Competency Assessment. The descriptors you use will depend on your elements and subelements, as well as your organizational structure.

Because competency in safety directly impacts the two primary Strategic Areas of Safety, performance and compliance, it is important to have a measure of the current level of competence, and incorporate competency improvement into your strategic safety plan, if warranted.

The ability of an organization to fulfill its mission and realize its vision depends on the capabilities of its people. The competence of safety professionals plays a critical role in fulfilling your safety mission and realizing your safety vision. Competence drives both your safety performance and compliance, and, as Potter said, is required for a sustainable safety culture. Therefore, assessing your organization's level of competency to determine if and what strategies are needed to improve it is a critical part of the strategic planning process.

The final chapter in Part III examines the analysis of culture, the fourth strategic area of safety. An organization's safety culture is the most important contributing factor in achieving safety excellence, so analyzing it is crucial to that effort.

Endnotes

1. Carl Potter, "Are You Competent When it Comes to Safety?" *EHS Today*, July 18, 2016, accessed October 29, 2023, https://www.ehstoday.com/safety-leadership/article/21917674/are-you-competent-when-it-comes-to-safety.

2. Ibid.

3. The Institution of Occupational Safety and Health, "Competency Analysis: Professional Standards for Safety and Health at Work," accessed November 4, 2023, https://iosh.com/guidance-and-resources/professionals/competency-framework.

4. IChemE Safety Centre, "Process Safety Competency – a Model," 2015, accessed November 4, 2023, https://opuskinetic.org/wp-content/uploads/2019/01/safety-centre-competency-oct15-min.pdf.

5. Ibid., 10.

Chapter Nine:
Culture Analysis

Perception surveys are invaluable in pinpointing what actions are needed to improve safety systems — Dan Peterson

Every organization has a safety culture, even if they do not manage or measure it. If your company doesn't manage it, analyzing your culture for the first time is crucial to determine its maturity and quality. Since achieving safety excellence demands a mature safety culture, due to culture driving the other three Strategic Areas of Safety, this is perhaps the most important area of the four to analyze.

This strategic area focuses on aspects of safety involved with cultivating and managing the safety culture. Analyzing this area helps to answer the questions: "How good is our culture, how good do we want our culture to be, and how can we improve it?"

Before discussing how to analyze your organization's safety culture, it helps to have a good working definition of it. The definition provided by the United Kingdom's (UK) Health and Safety Laboratory (HSL), offers a clear and concise definition. On their web page titled "Achieving Safety Culture Excellence," the HSL defines it as such: "Safety culture is a combination of the attitudes, values and perceptions that influence how something is actually done in the workplace, rather than how it should be done."[1]

Any discussion about safety culture should also include a discussion about safety climate. Safety culture is often used to refer to both, collectively, even though there are unique differences between the two terms. In this chapter, safety culture is used as a combination of both terms, except where safety climate is specifically addressed. It is important to know the difference between the two, however, before moving on to a discussion about assessing the safety culture.

Most definitions of safety climate refer to it as something local rather than organizational, and related to the current state instead of the desired state. In an IChemE paper titled "Toward a Mature Safety Culture," the paper's authors describe the differences between the terms as follows:

> Safety culture may therefore be defined in terms of underlying belief systems about safety which are partly determined by group values, norms and regulatory frameworks. Safety climate, on the other hand, refers to the state of a system in terms of perceptions of the current environment or prevailing conditions which impact upon safety.[2]

There are two popular assessments related to the level of maturity of your overall safety program and the state of your safety culture. They include:

1. *Maturity Assessment:* This assessment analyzes the maturity level of your overall safety program and/or safety culture as they progress through various stages of maturity. This assessment provides a snapshot of a point in time and helps you identify improvements needed to progress to a higher level of maturity. This assessment is crucial for any organization pursuing safety excellence.

2. *Perception Survey:* This assessment solicits feedback from the organization, including managers, supervisors, and front-line employees, using a carefully crafted questionnaire about the safety climate. The data generated from the survey is used to identify weaknesses in safety systems and mindsets, and to determine how to address them.

These two assessments offer an excellent source of insight into what objectives you should include in your safety strategy to achieve safety culture excellence. This chapter provides an introduction to these assessments and presents a practical approach for performing them. Once again, the primary objective is to help you see the value in performing them as part of your strategic planning process.

Maturity Assessment

A Maturity Assessment is similar to the Competency Assessment described in the previous chapter. They both involve a framework of elements that are assessed against a scale. With Maturity Assessments, the scale is based on level of maturity. The framework's elements in this case are used to assess an organization's overall safety program, as well as its safety culture.

Culture Framework

Most safety-related Maturity Assessments claim to measure the safety culture. Some are narrowly focused on cultural aspects of safety, while others incorporate safety management system aspects. The dss[+] Bradley Curve[TM], once treated as the gold standard of safety maturity models, is an example of a culture-based model.

A DuPont employee, Berlin Bradley, developed the model in 1995. It includes four stages of safety culture. It is based on the assumption that high injury rates are because of a poor safety culture, and therefore relies on employees' perception of safety to assess the maturity of the culture.

Terry L. Mathis, founder of ProAct Safety, wrote two articles in *EHS Today* critiquing that model and pointing out its shortfalls in achieving safety culture excellence. The first article, titled "Safety and Performance Excellence: Behind the Bradley Curve," provides more of a background on the model and what he believes are the flaws in its premise.[3] In the second article, titled "Safety Excellence Maturity Model," Mathis states organizations achieve true safety excellence by addressing four major and two minor elements, more culturally based than systems-based.

> The major elements are: strategy, leadership, employee engagement and culture. The two minor, but important, elements are the role of safety professionals and safety metrics. While safety culture is a major element, it is so strongly impacted by the other five that working on culture without addressing other elements is ineffective…The real issue is how to get the culture to progress through these stages and that involves strategically managing these other elements toward that goal.[4]

My reason for mentioning the Mathis articles is not to disparage the dss[+] Bradley Curve[TM], but to introduce the differing viewpoint that a broader range of elements more accurately assesses the safety culture. As designed, the dss[+] Bradley Curve[TM] lends itself more to a Safety Perception Survey, which is how they currently market it.

The Keil Centre is a consulting and training group in the United Kingdom specializing in psychology and related disciplines. It developed the Safety Culture Maturity ® Model. For comparison sake, that model includes the following ten elements, including both cultural and systems-based elements:[5]

1. Visible Management Commitment	6. Health and Safety Resources
2. Supervision	7. Competency
3. Learning Organization	8. Contractor Management
4. Production Versus Safety	9. Risk-Taking Behavior
5. Safety Communication	10. Participation In Safety

(Safety Culture Maturity is a Registered Trademark of The Keil Centre Ltd. Copyright The Keil Centre, 1999.)

The Safety Management Framework used in Chapter Eight, and described further in Chapter Ten, is used in this example for maintaining consistency and promoting integration of safety assessments.

There is another reason, however, that the Safety Management Framework is valid for this assessment, one directly related to the safety climate.

Chris Goulart was the Director of Safety Services for Regulatory Consultants, Inc. In an archived blog post on the *Occupational Health & Safety* website titled "Resolving the Safety Culture/Safety Climate Debate," Goulart explained the direct relationship between a safety management system and the safety climate.

> Since Safety Climate is a measure of what is occurring in the instant, it is directly influenced by events that have recently occurred. By direct extension, the views and perceptions about safety today are molded by the very recent activities of the organization. This means that the current elements of an organization's Safety Management System (SMS) have a direct impact on the Safety Climate. Thus, the metrics that are measured and evaluated by the SMS are foundational elements that make up the Safety Climate.[6]

The Safety Management Framework includes systems, cultural, and behavior-based elements. Collectively, they comprise all aspects of a safety program. Each aspect of it helps drive the safety climate.

Maturity Scale

Like a competency scale, a maturity scale is a set of descriptors and their definitions that follow a progression of maturity. Some maturity models use a four-point scale, but most use a five-point scale. The descriptors used differ among various models. Below is a brief review of three of the well-known models used to assess safety culture maturity.

The dss[+] Bradley Curve™, introduced above, includes four stages of safety culture representing different levels of employee responsibility for safety. This is more of a behavior-based safety association. The four stages are:[7]

1. *Reactive stage:* People do not take responsibility. They believe that safety is more a matter of luck than management and that incidents will happen.

2. *Dependent stage*: People view safety as a matter of following the rules that someone else makes. Incident rates decrease and management believes that safety could be managed if only people would follow the rules.

3. *Independent stage:* Individuals take responsibility for themselves. People believe that safety is personal, and that they can make a difference with their own actions. This reduces incidents further.

4. *Interdependent stage:* Teams of employees feel ownership for safety and take responsibility for themselves and others. People do not accept low standards and risk-taking. They actively converse with others to understand their point of view. They believe true improvement can only be achieved as a group, and that zero injuries is an attainable goal.

The Hudson Safety Maturity Model, referred to as the Hudson Ladder, was created by Patrick Hudson, Ph.D., Professor Emeritus at the Delft University of Technology in the Netherlands. Hudson is a psychologist and one of the world's leading authorities on human factors in safety in the oil and gas, commercial aviation, mining, and health care industries. His model, a maturity scale that defines five levels of maturity, was provided in a 2001 *Flight Safety Australia* article titled, "Safety culture: The ultimate goal." Below are the five levels of maturity.

1. *Pathological:* The organization cares less about safety than about not being caught.

2. *Reactive:* The organization looks for fixes to accidents and incidents after they happen.

3. *Calculative:* The organization has systems in place to manage hazards. However, the system is applied mechanically. Staff and management follow the procedures but do not necessarily believe those procedures are critically important to their jobs or the operation.

4. *Proactive:* The organization has systems in place to manage hazards and staff and management have begun to acquire beliefs that safety is genuinely worthwhile.

5. *Generative:* Safety behavior is fully integrated into everything the organization does. The value system associated with safety and safe working is fully internalized as beliefs, almost to the point of invisibility.[8]

The Safety Culture Maturity ® Model, also introduced above, uses a different five-point scale with descriptors more suitable for safety culture, though having some management systems undertones. Below are the five stages it uses:[9]

1. *Emerging:* Safety is not perceived as a significant business risk, and incidents are considered a part of doing business. Responsibility for safety resides within the safety department. Mistakes are blamed on the front-line employees.

2. *Managing:* Safety is perceived as a business risk, but one that can be solved with safety rules and procedures. There is a reactive response to incidents. Mistakes are blamed on the front-line employees, but management takes responsibility for safety performance. Lagging indicators are used to measure safety performance.

3. *Involving:* Management realizes that supervisors and front-line employees must be involved in safety, and personal responsibility is emphasized. The organization is tracking leading indicators.

4. *Cooperating:* Management takes proactive measures to prevent incidents. Employees understand the importance of safety and are engaged in promoting it. The organization uses leading indicators as measures of safety performance.

5. *Continually Improving:* Safety is a core value of the organization. The organization routinely reviews safety performance to ensure continuous improvement. Employees hold themselves accountable for both their own safety and the safety of others.

These three maturity scales should give you a clear idea of the stages that a safety culture can go through, both up and down the ladder, depending on how well the safety culture is managed.

There are maturity models based on an organization's capabilities rather than its culture. The KPI Institute has a series of maturity assessments for other aspects of the business besides safety. They use a set of five descriptors that are much more descriptive of management system maturity. These are: *initial, emergent, structured, integrated,* and *optimized.* These descriptors would also be helpful when assessing maturity using the Safety Management Framework.

Since the Safety Management Framework includes various elements related to safety management, it makes sense to have different descriptors for different aspects of it, along with descriptors for safety culture. Table 26 provides a generic five-point scale with three-tiered descriptors for assessing the Safety Management Framework's elements and sub-elements.

Table 26.   Generic Maturity Scale

Aspect	Level/Stage 1	Level/Stage 2	Level/Stage 3	Level/Stage 4	Level/Stage 5
Management Systems	**Emergent**	**Developed**	**Coordinated**	**Integrated**	**Optimized**
	Management systems are in the early stages of development and Implementation.	Management systems are developed, individually, and are being Implemented.	Management systems are standardized and coordinated with each other.	Management systems are integrated with other business management systems.	Safety performance is analyzed and used to develop strategies for improvement.
Incident Prevention	**Dismissive**	**Reactive**	**Responsive**	**Proactive**	**Predictive**
	Management dismisses all but the most serious incidents as being unavoidable, and typically blame those involved.	Management reacts to incidents after they occur and take corrective actions to address unsafe acts or conditions.	Management responds to incidents in a timely manner, taking both corrective and preventive actions.	Management is actively involved in identifying ways to control hazards and reduce risk even when incidents do not occur.	Management tracks, trends, and analyzes incidents and leading indicators to predict and prevent future incidents.
Personal Responsibility	**Blaming**	**Complying**	**Participating**	**Collaborating**	**Leading**
	Management places the responsibility for safety on the safety staff, and no one else is held accountable. Employees blame poor safety performance on management..	Management starts assuming responsibility for safety performance. Employees comply with safety rules and procedures, but do not assume responsibility for performance.	Management engages and supports supervisors as the first line of defense in safety. Employees participate in the safety process and share responsibility for safety performance.	Management works with all levels of the organization to manage and improve safety. Employees work with management to improve safety, and take personal responsibility for their own safety.	Management develops and supports safety leaders at all levels. Employees start leading safety, and take personal responsibility for their own safety and the safety of others.

Another approach to describing the maturity scale is by adding definitions for each of the Safety Management Framework's elements. This provides a greater level of detail and provides for a more accurate assessment.

You must take care in writing the element descriptors. If the descriptors do not describe the level well enough, you will have a harder time rating each level. Also, when writing your descriptors, ensure that they truly reflect a progression through the levels.

Table 27 provides an example of a specific maturity scale for the second set of four elements of the Safety Management Framework, the execution elements. Creating a generic maturity scale first helps to guide you in writing specific definitions for each element that progress consistently throughout the levels or stages.

Table 27.   Specific Maturity Scale

Element	Levels/Stages				
	Emergent/ Dismissive/ Blaming	Developed/ Reactive/ Complying	Coordinated Responsive/ Participating	Integrated/ Proactive/ Collaborating	Optimized/ Predictive/ Leading
Contractor Management	Contractors are being hired without regard for safety, other than contractual requirements to comply with OSHA.	Contractor Safety Program is in place, and safety orientations are being performed prior to work commencing.	Safety programs are being shared between the site and contractors, and a safety pre-qualification process is in place.	Contractor safety observations and post-work evaluations are being conducted. Contractor injury rates are tracked.	Contractor safety performance is analyzed and used to develop strategies for improvement.
Risk Management	Risk Assessments & Method Statements are developed for field service work when required by customers.	A Risk Profile has been completed, as have any Regulatory Assessments and Surveys required by OSHA regulations.	Hazard Assessment and JSA processes exist, and are coordinated with applicable safety programs.	Ergonomic Analyses are being performed and all risk assessment processes are fully integrated with other safety processes.	All risk assessments are periodically reviewed and revised, and any gaps are taken into consideration for strategic planning.
Safe Work Processes	Safe Work Procedures (LOTO, CSE, etc.) are being developed and implemented to comply with OSHA regulations.	Software has been implemented to manage safe work processes. Stop Work Authority has been implemented.	Life Saving Rules are in place and coordinated with all of the safety processes, as well as the disciplinary policy.	A work Authorization process and combined Safe Work Permit are implemented to integrate all safe work processes.	Safe Work Permits are periodically reviewed and revised, and any gaps are taken into consideration for strategic planning.
Safety Work Behaviors	Employees blame poor safety performance on management, and take little or no personal responsibility for it.	Employees comply with safety rules and procedures, but do not assume responsibility for the outcome.	Employees participate in the safety process, perform safety observations, and pursue housekeeping excellence.	Employees work with management to improve safety, and take personal responsibility for their own safety.	Employees start leading safety, and take personal responsibility for their own safety and the safety of others.

Since it is a progression, a maturity scale implies that you cannot jump or skip a level. You cannot jump straight from childhood to adulthood; you must progress through the various stages of life, contrary to what most teenagers might think. You gain new knowledge and experience at each stage that prepare you for the next. The same applies to safety management systems and culture.

Patrick Hudson gave a talk titled "Safety Culture and Leadership," during which he explained the need to progress through each stage of culture maturity. He advised against trying to shoot for a higher level without being firmly planted in the one below it.

I've come to realise that safety leadership, at least its content rather than its style, can be framed in terms of going one step higher than those around you. So, if the culture is reactive, then leadership involves setting out a calculative vision and behaving in ways that make that come about. If the culture is already calculative, then the leader has to be dragging everyone else up towards the proactive stage, and if the culture is still predominantly pathological, and many unfortunately are, then reactive is where you have to be. Shooting for proactive is doomed to fail. No one will understand what you say and they lack the skills to build on it.[10]

Hudson also described the importance of safety leadership developing a vision for the next higher level of maturity, before pursuing it.

In all these cases, safety leadership appears to have involved the leader identifying and publicly pushing a vision that represented one step higher on the safety culture ladder than the organization was at the time. The advantage of using this knowledge is that a safety leader can articulate for themselves a much more detailed vision of what they aspire to than just being safe or safety first. Anyone armed with a clear and more detailed vision–what, how and why–will find it much easier to communicate, to know how to behave whatever their personal leadership style. These three–what we do, how we do it and why–differ at the different levels of the safety ladder, changing as we learn to get better and progress.[11]

The element descriptors for each level of maturity in tables 26 and 27 serve as a part of the visions described by Hudson, visions that help you progress through the levels of maturity, one level at a time.

Assessing Culture

As with other assessments, it is better to perform culture assessments with a representative group of people, including those at different levels within the organization and performing various functions. Safety cultures and the perception of them can vary throughout the organization.

You can score the assessment in a couple of ways. The first way simply determines the level of maturity each element is at by comparing it to the descriptors and their definition. The second way—my preference—is using a percentage of achievement for each element at each level. Table 28 provides an example of a maturity assessment of the Safety Management Framework execution elements following this approach. It uses a scale with 20% increments in the scoring method.

Table 28.   Safety Management Framework Maturity - Execution Elements

Element	Sub-Elements	Percent Achieved by Level/Stage					Maturity Level
		1	2	3	4	5	
(Portion of Table Omitted)							
Contractor Management	Contractor Safety Program	100%	100%	100%	80%	20%	3.8
	Contractor Safety Pre-Qualifications						
	Contractor Safety Orientation						
	Contractor Safety Observations						
	Contractor Safety Evaluations						
Risk Management	Risk Profile (Risk Register)	100%	100%	40%	0%	0%	2.4
	Regulatory Assessments & Surveys						
	Risk Assessments & Method Statements						
	Hazard Assessments & JSAs						
	Ergonomic Analysis (REBA, RULA, etc.)						
Safe Work Processes	Work Authorization Process	100%	100%	80%	20%	0%	2.8
	Safe Work Permits						
	Safe Work Procedures (LOTO, CSE, etc.)						
	Life Saving Rules						
	Stop Work Authority						
Safety Work Behaviors	Personal Responsibility for Safety	100%	100%	40%	0%	0%	2.4
	Safety Observation Program						
	Adherence to Safety Rules						
	Proper Use of PPE						
	Housekeeping Excellence						
Maturity Level	Average Percent Achieved	100%	100%	65%	25%	5%	
	Overall Maturity Level	2.65					

As shown in table 28, lower levels are 100% achieved, whereas higher levels have not been, although some progress is possible at all levels. Each element has its own maturity level computed based on the percentage achieved at the various levels. Looking at the contractor management element the maturity level of 3.8 reflects the fact that levels one through three were 100% achieved, and level four was only 80% achieved. (Nested IF AND formulas were used to generate the value.) The same method was used to determine the overall maturity level achieved, based on the average percentage achieved for each of the five levels. This would also be done for the entire framework.

The other metric is the maturity level for each element, which tells you where improvement is needed. In table 28, the elements most in need of work are risk management and safe work behaviors. However, safe work procedures also needs work to get it up to 100% for level three. Having already achieved level three in contractor management, no short-term improvement is needed in that element.

Table 29 provides an example of a maturity assessment of the final four elements of the Safety Management Framework, the examination elements. Between table 24 (Chapter Eight), table 28 and table 29, you can see all twelve elements of my framework, and the sixty sub-elements. If you wanted to just assess those elements directly related to safety culture, I suggest assessing the visible leadership, culture and communications, safety work behaviors, and incident management elements.

Table 29.   Safety Management Framework Maturity - Examination Elements

Element	Sub-Elements	Percent Achieved by Level/Stage					Maturity Level
		1	2	3	4	5	
Audits & Inspections	Safety Management System Audits	100%	100%	100%	60%	20%	3.6
	Compliance & Program Audits						
	OSHA Mandated Audits & Evaluations						
	Site Safety Inspections						
	Safety Equipment Inspections						
Change Management	Management of Change Procedures	100%	100%	80%	20%	0%	2.8
	Facilities & Equipment Change Mgmt.						
	Process & Engineering Change Mgmt.						
	Management Systems Change Mgmt.						
	Organizational Change Mgmt.						
Incident Management	Crisis Management Policy	100%	100%	40%	20%	0%	2.4
	Emergency Preparedness & Response						
	Incident Investigation & RCA						
	Case Management & Return to Work						
	Business Continuity Plan						
Analysis & Strategy	Periodic Management Reviews	100%	100%	80%	40%	20%	2.8
	Performance & Compliance Analysis						
	Competence & Culture Analysis						
	Strategic Planning Process						
	Strategic Safety Plan & Execution						
Maturity Level	Average Percent Achieved	100%	100%	75%	25%	5%	
	Overall Maturity Level	2.75					

This assessment provides two valuable metrics for strategic planning. The first is the overall maturity level achieved. In strategy terms, this answers the question, "Where are we?" It also tells you where you want to go, which is to fully achieve level three before working to achieve level four.

A critical question when using Maturity Assessments for strategic planning is: "How long does it take to achieve a mature safety culture?" According to experts, it takes one to two years for the culture to be firmly established at each level. Therefore, using a five-point scale, it should take from five to ten years to progress to the highest level if your organization is new or still not firmly established at the lowest level. For established organizations, you will likely start at a higher level when you begin your journey.

Maturity Assessments are a great way to analyze your safety management system and/or safety culture for strategic planning. It begins with having the right framework and maturity scale for your organization.

Safety Perception Survey

The organization's values, beliefs, and attitudes regarding safety shape the design and implementation of its safety management systems, as well as its safety activities and behaviors. Its safety climate encompasses the organization's collective perception of the current state of its safety systems, activities, behaviors, and their impact on safety. Therefore, the safety climate is a barometer of the organization's safety culture as exhibited in day-to-day safety operations. Safety Perception Surveys are the most common way of obtaining data about the safety climate.

In his book, "*Safety Management: A Human Approach, Third Edition*," Dan Petersen describes the value of Safety Perception Surveys. He states:

> After many years of development and testing, it has been determined that such surveys are a much better predictor of a company's future safety performance than other indicators tested. Perception surveys are invaluable in pinpointing what actions are needed to improve safety systems.[12]

Below we will review the key issues pertaining to the use of perception surveys to provide a strategic perspective of your safety culture.

Survey Content

Developing survey content is a critical part of the survey process. Commercial surveys are quicker to implement, but the cost may be prohibitive.

Below is a brief description of five commercial Safety Perception Surveys offered by national safety organizations and reputable consulting firms, each of which offers consulting services to support the survey process.

1. *NSC Safety Barometer Survey:* The National Safety Council (NSC) offers this survey. The standard version is fifty questions, though site-specific questions can be added. It takes approximately fifteen minutes to complete and is available in electronic and hard copy formats. It also offers benchmarking against other organizations in the same industry.[13]

2. *Safety Climate Tool (SCT):* The Health and Safety Executive (HSE) in the United Kingdom (U.K.) offers this survey. The HSE provides a wide variety of support materials for promoting and conducting the survey project. They administer the survey electronically, and it offers benchmarking against other organizations in the same industry.

3. *The dss⁺ Safety Perception Survey™:* dss+, an independent consulting firm that spun off from DuPont in 2019 as DuPont Sustainability Solutions (DSS) offers this survey. It was rebranded in 2022. The survey results can be benchmarked against other organizations in the same industry.[14]

4. *Culture Change Consultants (CCC):* This consulting firm, founded by Steven I. Simon, Ph.D., the father of safety culture, offers the CCC Safety Culture Perception Survey. It is a fifty-two-item survey tested for reliability and validity.[15]

5. *ProAct Safety:* This consulting firm, founded by Terry L. Mathis, offers a Safety Perception Survey that can be used out-of-the-box or customized for your organization, the latter being the recommended approach.[16]

If you performed a survey in the past, you should repeat that survey to collect the same data for verification and trend analysis purposes.

Safety Perception Surveys are comprised of questions and/or statements that are grouped under various elements of a safety management system or culture. These elements make the resulting data meaningful for strategic planning, provided they align with your organization's safety systems.

As an example, the NSC survey measures six safety management system elements, similar to other perception surveys, including: "management leadership and commitment, supervisory engagement, employee involvement, safety support activities, safety support climate and organizational climate."[17]

The Culture Change Consultants survey aligns with the firm's Simon Open System (S.O.S.) Culture Change Model™. "The twelve organizational and cultural dimensions that are measured are leadership, rituals, values, norms, rewards, measurements, structure, social processes, technology, environment, heroes and symbols."[18]

The number of survey questions is important to the survey's effectiveness and level of participation. If your survey has too few items, it will not provide enough data for an accurate measure of the safety climate. Too many items will either cause respondents to rush through it and provide inaccurate responses, or avoid completing it altogether.

There is no magic formula for the ideal length of a survey, but one common rule of thumb offers some guidance. For each element measured, you should have a minimum of three to five items to ensure it is adequately addressed. The NSC survey includes fifty items with six elements, resulting in an average of eight and one-third items per element. The Culture Change Consultants survey has fifty-two items and twelve "dimensions," resulting in an average of four and one-third items per dimension.

A survey containing twenty-five element-based items might be a good minimum length, although a fifty-item survey would allow for reverse coding like items. (Reverse coding is explained later.) If a fifty-item survey takes fifteen minutes to complete, as the NSC survey suggests, a one-hundred item survey would take approximately thirty minutes to complete. This would probably push the limits of a respondent's attention span or interest. They may get survey fatigue and not complete the full survey.

That is one metric associated with surveys, *completion rate*. It is the percentage of respondents who fully finish a survey after starting it.

(Eq. 3) *Completion Rate = (# of Fully Completed Surveys / # of Surveys Started) x 100*

A low completion rate means respondents are not providing all of the data you want, and you want all of the data you can get. A low completion rate can be a sign of poorly written content, resulting in respondents skipping over items, or survey fatigue.

Keep in mind that the survey must also include demographic items, such as job title, shift, location, etc. However, never ask for a respondent's name! Keep the survey strictly anonymous. One key use of demographic information is to separate results for managers and supervisors, and front-line employees in order to measure perception gaps, as described below. Demographic items are best added to the end of the survey to avoid raising doubts regarding the anonymity and confidentiality of responses.

The quality of survey items is extremely important. The surveys listed above were professionally developed and tested for reliability and validity, the two measures of quality for organizational surveys.

1. *Reliability:* Reliability refers to the consistency of results, how well the survey produces similar results under the same conditions. For example, survey re-

sults for a facility should produce relatively consistent responses among employees, since it is the same safety climate being measured. Likewise, the same employee taking the survey a few weeks apart should provide similar results.

2. *Validity:* Validity refers to the accuracy of the results, how well the survey measures the intended content. For example, survey results should align with the safety performance of the organization or facility surveyed, since safety culture drives safety performance. If the survey results in higher scores, but the safety performance was poor, the validity of the survey would be questionable. How could a business with such a good safety culture perform so poorly?

Some organizations develop their own perception survey, with or without professional support. If you plan to develop your own survey based on your expertise in safety, you should also consider your level of knowledge in organizational surveys. You should conduct thorough research on this topic before creating your own survey. What follows should give you a good start.

Although sample survey questions or statements are available online, you need to carefully consider the wording and structure of each question or statement. Below is a description of key factors to consider regarding survey items to help ensure the quality of your survey.

Open-Ended versus Closed-Ended

Open-ended or write-in questions require an unstructured textual response that provides detailed qualitative data, which reveals a deeper understanding of the respondent's opinion. Closed-ended questions or statements provide for limited responses that may include binary data (e.g. true/false or yes/no) or ordinal data such as a Likert scale. You can statistically analyze data from closed-ended questions but not open-ended ones.

Most surveys use closed-ended items, although some open-ended items are recommended by experts. Another effective option is to allow for optional open-ended comments following each closed-ended item, providing both analytical and descriptive data. I used this approach on a multinational Safety Perception Survey and found the added comments very helpful in interpreting the results.

Response Scale

Most perception surveys use a Likert scale for responses to closed-ended survey items. A Likert scale is a response rating scale using five or seven levels, most commonly five. Some surveys use the same scale throughout the survey, whereas others use different scales based on the survey items.

Table 30 provides a sample list of five-point Likert scales based on the response type needed for different survey items.

Table 30.   Sample Likert Scales

Response Type	Response Scale				
	1	2	3	4	5
Agreement	Strongly Disagree	Disagree	Neutral	Agree	Strongly Agree
Comparison	Well Below Average	Below Average	Average	Above Average	Well Above Average
Satisfaction	Very Dissatisfied	Dissatisfied	Neutral	Satisfied	Very Satisfied
Effectiveness	Very Ineffective	Ineffective	Neutral	Effective	Very Effective
Quality	Very Poor	Poor	Average	Good	Very Good
Extent	To No Extent	To a Small Extent	To a Moderate Extent	To a Great Extent	To a Very Great Extent

Clear & Concise

Respondents must be able to understand precisely what is being asked. Survey items must be clear and concise to avoid misinterpretation. Consider the following when creating survey items:

- Do not use words signifying absolutes, like "always, never, all, or none," since respondents may question how to rank responses.

- Using double negatives in a survey item also adds confusion. Here is an example, "Do you not disagree that the company provides adequate resources for safety?" Respondents would likely not know whether they should answer if they agree or disagree.

- The following double-barreled question asks about two different things, "Do you receive safety training and is it effective?" Respondents are left to guess which one to rate, and might do so differently.

Also, ensure that you use proper grammar to avoid misunderstanding.

Familiarity & Readability

Survey items should include language familiar to respondents. If translating a survey into multiple languages, ensure the translators use the current local vernacular. Do not rely on simple online translators.

The readability of your questions and statements is also important. Avoid technical terms, complex words or sentences, and uncommon abbreviations or acronyms. Do not assume respondents will have the same level of knowledge that you do.

Leading & Loaded Questions

Do not use leading questions or statements that influence or manipulate respondents to give a particular response. The question, "Do you agree that our company has the best safety program in the industry?" is a leading question that could influence respondents to respond positively on an agreement scale. A more neutral question would be, "How does our organization's safety program compare to others you are familiar with?"

Loaded questions or statements are those that include unverified or unjustified assumptions that prevent a respondent from being able to respond accurately. The statement, "Your supervisor disciplines you when you intentionally break the safety rules," assumes that the respondent intentionally breaks the rules. A respondent cannot answer that accurately if he or she does not intentionally break the rules.

Reverse Coding

Survey items can be written in the positive or negative. "Management recognizes and rewards good safety efforts," is stated positively, whereas "Management does not recognize and reward good safety efforts," is stated negatively. Reverse coding is phrasing questions or statements in the opposite polarity.

Acquiescence or agreement bias is a common survey bias. This is the tendency to answer more positively to acquiesce to the norm or to make themselves appear more agreeable. You may have noticed this tendency when completing a survey following a seminar, finding yourself to be more complimentary in your course evaluation responses.

Respondents can also get into a pattern of answering in the positive over several items and then pay less attention to the remaining survey items.

Reverse coding, or phrasing some items in the negative, in the case above, helps to minimize the potential for agreement bias, and also requires respondents to read survey items more carefully to confirm its polarity.

Reverse coding has another use, stating the same survey item in both the positive and negative terms to analyze how consistent a respondent was in their responses. If a strongly agree response was given for the item positively phrased, you would expect a strongly disagree response for the same item negatively phrased. A strongly agree response for both would show that the respondent was not giving accurate responses.

The factors listed above, as well as the survey response rate described below, impact a survey's reliability and validity. If you create your own surey, know thi:

- A poorly written survey is likely to cause respondents to select responses randomly or even abandon the survey altogether.

- Too many unclear or wordy survey items make it more burdensome for respondents to complete the survey, making it more mentally taxing.

- Poorly translated surveys can confuse respondents or even insult them by appearing disrespectful to their native language.

- Leading questions give respondents the feeling that the survey has an ulterior motive, such as making the company look good as opposed to seeking genuine feedback.

Given the time and resources that go into implementing a Safety Perception Survey, it pays to develop quality survey content. Otherwise, it could end up being a waste of valuable resources.

Survey Administration

Safety Perception Surveys require advanced planning, and a sufficient amount of time needs to be allotted for each phase of the project. The overall project could take from two to six months, depending on the circumstances. The target date for the survey provides a starting point to work back from for project planning purposes.

Below is a brief description of the seven phases involved in implementing a Safety Perception Survey, and some of the key considerations to ensure its success. Time estimates are provided, but should be considered in terms of your unique circumstances.

Phase 1 - Survey Planning

The initial planning begins with building the business case for running a survey, and getting leadership approval and commitment. This may require establishing the objectives and benefits of the survey, choosing the course of action for implementing it, including whether to use a commercially available or custom-developed survey, and determining whether to use consultants. This step could take two weeks to two months.

Phase 2 - Survey Development

This step involves either developing a custom survey or working with the survey vendor to customize and brand their survey for use by the organization. It also includes translating the survey, if needed, and formatting the survey for electronic and/or hard copy distribution. Depending on the distribution method, you need resources to collect and tabulate the responses. Developing the promotional material for the survey must also be considered. This could take a few weeks to a few months to complete.

Phase 3 - Survey Pilot & Testing

In larger organizations with multiple locations, it is advisable to run a pilot of the survey at one or more locations to identify any improvements needed, and to work out any logistical issues. Given the limited distribution, this could take about a month, followed by another week or two to evaluate the results and revise the survey, if needed. A survey pilot is especially helpful if your survey was developed in-house and is the first survey implemented.

Phase 4 - Survey Promotion

For Safety Perception Surveys, promotion prompts participation, participation produces data, and data generates insight. The promotion step, performed just prior to and while administering the survey, involves publicizing and promoting the survey to maximize participation and quality of responses. You should treat the publicity and promotion as a marketing campaign. Instead of generating customers, its purpose is to generate participants. Like a marketing campaign, it should be multimedia, incorporating as many forms of communication as needed.

The higher the participation level is, the more accurate the results are. To help ensure that people will take part, the promotion should include the following crucial elements:

- Objectives, uses, and benefits of the survey.

- Support and commitment of senior leadership, and the value they place on employee feedback.

- Assurances of anonymity and confidentiality.

- Means of accessing and submitting the survey.

An important question is, "What level of participation should we shoot for?" Participation rate (a.k.a. *response rate)* is calculated as follows:

(Eq. 4) *Response Rate = (# of Completed Surveys / # of Employees Invited) × 100*

There are three response rates to consider: the desired response rate, the acceptable response rate, and the actual response rate. Before starting, you should have a good sense of what the first two rates are.

While many argue that 100% participation is ideal, others do not. Culture Amp, a leading employee experience platform, argues that 100% participation will include responses from employees who may have felt coerced into participating and likely provided incomplete or skewed data. On their webpage titled "What is a good employee survey response rate?" Culture Amp states: "In that sense, having a 100% response rate is sometimes associated with poorer quality data–and it is the quality of data that we should be focused on, not the quantity."[19]

Culture Amp describes desirable response rates as follows:

Generally, participation rates for staff surveys fall between the 65-85% range. These are rates that political scientists and market researchers can only dream of…

In small companies or teams (less than 50), we should be aiming a little higher than the 65-85% range. At Culture Amp, we usually recommend an 80-90% rate to be a good minimum benchmark, as it allows us to hear from 4 out of 5 people on average.

As we move to larger companies, we can scale our expectations down. With 500 employees, we will probably get a good sense of where the company is with a 70% rate of participation, so 70-80% is a good benchmark.

Companies of 1000+ can probably aim for a participation rate of around 65% as their lower bound–even though higher rates allow a stronger sense of involvement psychologically.[20]

A key consideration when determining your desired and acceptable response rate is the margin of error (MoE), and its associated confidence level. The MoE measures uncertainty. It represents the maximum expected difference between the theoretical total population result and your survey result. A confidence level of 95% is the standard level used.

The smaller the MoE you want, the larger your response rate will need to be. SurveyMonkey[R] is the world's largest online survey platform, which I have used successfully for Safety Perception Surveys. In an online article titled "What is a good survey response rate?" they provide a convenient table showing the number of responses needed in various populations to yield MoEs of 3%, 5%, and 10%.[21] They also provide an MoE calculator and instructions on how to calculate it.

What MoE should you shoot for? SurveyMonkey[R] states in the article that: "The level of tolerance for inaccuracy will depend on your confidence in being able to make decisions based on the data you obtain, and of course, on the importance of the decisions you're making."[22] Generally speaking, a ± 3% MoE is desirable, and a ± 5% MoE is acceptable, based on the standard 95% confidence level.

Using Culture Amp's recommended response rate of 70-80% for a population of 500, SurveyMonkey's[R] table shows that a 69% response rate (345/500) has a ± 3% MoE. For an MoE of 5%, only a 44% response rate is needed. For a population of

1,000, a response rate of 52.5% is needed for a $\pm$ 3% MoE, and only 28.5% for an MoE of 5%.

Therein lies the importance of good survey publicity and promotion, maximizing the participation so you obtain your desired response rate, producing results with the level of accuracy you want or need.

The promotion must run long enough to ensure that the entire workforce gets the word, and to motivate them to participate in the survey. This could take at least a few weeks.

Phase 5 - Survey Implementation

Survey implementation is another key to maximizing participation. The three critical considerations in this phase are: *timing*, *technology*, and *tracking*.

1. *Timing:* You want to deliver the survey when the availability of participants is greatest. For starters, give employees time during their workday to complete it, and never expect them to complete it on their own time. Avoid delivering the survey during peak operating periods so employees have adequate time to complete it without feeling pressured by the demands of production. Avoid high vacation periods.

Allow a few weeks to a month for employees to complete the survey. Shorter periods might miss employees out on personal time. Giving too much time increases the risk of procrastinators forgetting the survey altogether.

2. *Technology:* The technology used to execute the survey needs to be considered, minimizing logistical and participation issues. Electronic surveys are easier to deliver and collect data from. If you use electronic surveys, the ability to deliver a survey on multiple devices will help improve participation rates.

Hard copy surveys are more difficult to collect data from. If you use hard copy surveys, I highly recommend that you use a scannable survey form to avoid having to count and transfer survey responses manually. Not only is that extremely time-consuming, especially for larger survey populations, it can also introduce errors during the transfer process, which would reduce the accuracy of your data.

There are two primary methods of executing a hard copy survey that will allow you to collect your data more easily by scanning survey forms.

2.a. *Scantron*^{TM} *Forms & Readers:* Scantron^{TM} forms and similar brands use Optical Mark Recognition (OMR) technology to read completed survey forms. The forms suit Likert scale responses well. Respondents simply need to color in the bubble for their chosen response. Survey forms are easily scanned using special OMR scanners.

2.b. Office Forms & Scanners: There are a variety of software solutions that allow you to create custom survey forms, using standard office software, that are scannable using office scanners. They use both Optical Mark Recognition (OMR) and Optical Character Recognition (OCR) technology. The latter allows for scanning textual data.

Depending on your organization, you might need to use both electronic and hard copy surveys to accommodate all potential respondents.

3. *Tracking:* During the implementation phase, track the progress of the survey participation rate. This helps to ensure that you achieve your desired rate by the end of the survey period. Tracking the overall participation rate, as well as rates by location and/or division, lets you see where your survey stands, and where you might need to focus your ongoing promotional efforts.

Phase 6 - Survey Analysis

At the end of the implementation period, collect and combine the data (i.e. electronic and hard copy surveys) into a single data set. If your organization has performed surveys before, you will analyze both current and historical data sets, separately. Do not combine historical and current data. Compare the two results to determine effectiveness of previous actions and to identify positive or negative trends.

There are four key metrics that need to be determined and analyzed to get the most benefit from your survey. They include: *participation rate, completion rate, percent positive*, and the *perception gap*. These metrics should be determined for each organizational level (i.e. corporate, divisional, departmental) and location (i.e. country, region, facility) so they can be analyzed separately to properly assess the local safety climates.

Participation and completion rates were described above. The other two metrics are described below using a hypothetical Safety Perception Survey comprised of 25 questions that are grouped into six categories. The survey uses a five-point Likert scale, and includes both positively and some negatively worded (i.e. reverse-coded) questions or statements.

Percent Positive

The percent positive score is a measure of how positive the responses were overall. The metric uses the Likert scale values and number of responses to determine the total points for each question, and compares it to the maximum points possible. The maximum score is based on all responses having a response value of 5, for five-point scales.

Table 31 shows how the percent positive score is calculated for a positively worded question or statement.

Table 31.   Percent Positive - Positively Worded Question

Front-Line Employees	# Of Respondents (A)	Point Value (B)	Total Points (A x B)	Maximum Points (A x 5)	% Positive
Always or Almost Always	7	5	35		
Often	14	4	56		
Sometimes	13	3	39		(Total Points / Maximum Points X 100)
Seldom	9	2	18		
Never or Rarely	2	1	2		
Subtotals	**45**		**150**	**225**	**67%**
Managers/Supervisors	# Of Respondents (A)	Point Value (B)	Total Points (A x B)	Maximum Points (A x 5)	% Positive
Always or Almost Always	15	5	75		
Often	10	4	40		
Sometimes	8	3	24		(Total Points / Maximum Points X 100)
Seldom	3	2	6		
Never or Rarely	1	1	1		
Subtotals	**37**		**146**	**185**	**79%**
Combined Totals	**82**		**296**	**410**	**72.2%**

As shown, the formula for the percent positive score is:

(Eq. 5)   *Percent Positive = (Total Points / Maximum Points) X 100*

First calculate the percent positive score separately for front-line employees and managers/supervisors. Then, combine these scores to calculate the overall percent positive scores. The individual scores are used later to calculate the perception gap between the two groups.

Table 32 shows how the percent positive score is calculated for the same question or statement worded negatively.

Table 32.   Percent Positive - Negatively Worded Question

Always or Almost Always	2	1	2		
Often	7	2	14		(Total Points / Maximum Points X 100)
Sometimes	15	3	45		
Seldom	13	4	52		
Never or Rarely	8	5	40		
Subtotals	**45**		**153**	**225**	**68%**
Always or Almost Always	1	1	1		
Often	3	2	6		(Total Points / Maximum Points X 100)
Sometimes	8	3	24		
Seldom	11	4	44		
Never or Rarely	14	5	70		
Subtotals	**37**		**145**	**185**	**78%**

Calculating percent positive scores for positively and negatively worded items is the same; they just require inverted Likert Scale point values. In the example above, both the positively and negatively worded items addressed the same issue. Therefore, we can compare the percent positive scores for the two to see how closely they align.

The percent positive score for the positively worded item was 72.68%, compared to 72.20% for the negatively worded itemt. That is a very close overall score. This shows that respondents paid careful attention to the questions to get the polarity right and answer accurately, which would otherwise diminish its validity.

The percent positive score is used to identify areas where the organization is performing well and areas that need improvement. A common threshold for determining areas needing improvement is a percent positive score of 80% or less. These are the areas you should evaluate further to determine if improvement actions are needed.

Figure 9 provides the percent positive scores for each category. Accountability, behavioral reinforcement, and management commitment are the areas that warrant further evaluation. That evaluation would include examining the percent positive scores of the underlying questions or statements, as well as any corresponding textual responses.

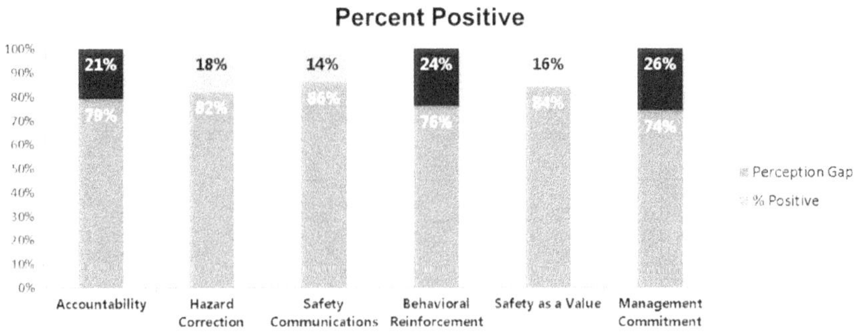

Figure 9.   Percent Positive Score by Category

Perception Gap

Historical Safety Perception Survey data shows that managers and supervisors often have a more positive perception of an organization's climate than that of front-line employees. Ideally, everyone in the organization should have a similar perception. Frequently, there are disparities between the perceptions of the two groups.

The perception gap is determined by comparing the percent positive scores for the two groups. It is calculated by subtracting the front-line employee score from the manager and supervisor score, the latter typically being the higher score.

Figure 10 provides the perception gap scores for each category. A gap or disparity of less than 10% is acceptable, whereas a gap of 10% or higher suggests the need for further evaluation, and possible corrective action.

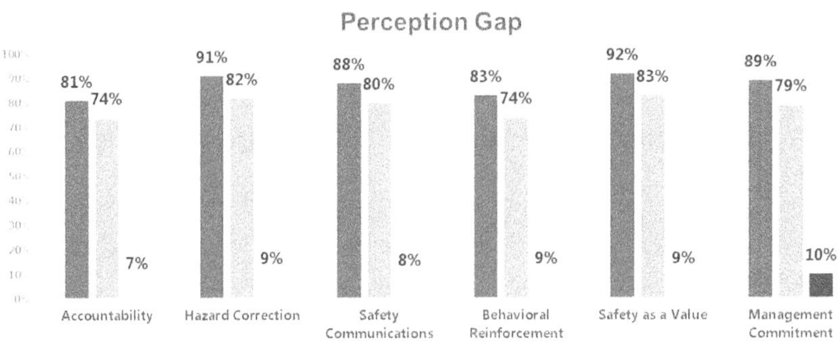

Figure 10.   Perception Gap Score by Category

The percent positive and perception gap scores provide valuable insight into the current state of the safety climate, and sheds light on those areas warranting improvement strategies. Not surprising is the fact that management commitment had the highest disparity in this example. If it was a follow-up survey, analyzing the current data against the historical data will show whether past action plans were effective, and reveal any positive or negative trends.

Phase 7 - Survey Follow-Up

Chapter Two states that you need specific targets and measures of success to determine if strategic objectives are achieved. This is true for strategic objectives developed from your Safety Perception Survey. You need a follow-up survey to determine whether you have achieved those objectives. Safety climates change over time for various reasons. Therefore, Safety Perception Surveys need to be performed periodically. The important question is, "How frequently should we perform them?"

Organizations should perform Safety Perception Surveys every one to three years. Performing them more frequently may not be an efficient use of organizational resources, since there would likely be insufficient time for improvement actions to bear fruit. Performing them less frequently may allow new negative cultural influences to go undetected for too long. One to three years is the sweet spot —I prefer two —for giving actions sufficient time to produce results, and not allowing negative changes to go undetected for too long.

For reference purposes, on the webpage introducing their safety perception survey brochure, dss+ recommends performing their survey "every 2-3 years to monitor and inform action plans."[23]

Some organizational changes can negatively affect the safety culture, such as changes in leadership at various levels, so it is better to catch those earlier rather than later. Longer periods between surveys make it more difficult to determine which underlying event(s) or change(s) drove a negative change in the safety culture. Also, many safety culture experts say it takes three to five years to improve a safety culture. Waiting until the end of that period would not allow for making timely course corrections, if needed.

I mentioned having used SurveyMonkey[R] for an international Safety Perception Survey in multiple languages, seven to be exact. The survey included business units in the U.S., U.K., Mexico, South America, Europe, the Middle East and Africa, and the Asia Pacific. Given the multiple languages and various types of operations, we didn't have adequate resources to support a large, multi-faceted campaign. Fortunately, we still had a response rate of 35%, which, for the size of our organization, resulted in a margin of error less than 3% based on a 95% confidence level.

It was administered both electronically and via hardcopy to accommodate employees without ready access to suitable devices. As for data collection, the data obtained from surveys taken on SurveyMonkey[R] was extremely easy to download. It consisted of a few thousand surveys, and I downloaded and entered the data into my analysis spreadsheet within just a few hours.

The nightmare came from the hardcopy surveys. Given our limited resources, we did not have the ability to scan survey forms using OMR or OCR readers, so that data was collected manually. Safety managers from the various business units simply scanned completed forms into large PDF files and sent those to me.

Imagine having to manually enter results from over 500 respondents for a twenty-five question survey. That's over 12,500 data points to enter into a spreadsheet, not to mention the textual comments, which I did mostly during my evening hours. It took me several nights to get that data entered, and those were some late nights!

Two aspects of the survey that worked very well were the demographic items and the open comments under each question. I was able to analyze the data not only for each business unit, but also by region, division, operation type, and the two companies involved in the joint venture. Due to the open comments, I was able to interpret the results more effectively, and develop better strategies for improving the culture in each segment of the business. Other than the manual data entry, it was a great experience.

An organization's safety culture has the greatest impact on its safety performance and compliance. Every organization has a safety culture, good, bad, or indifferent, regardless of whether they actively manage their safety culture. You can assess your organization's safety culture, including its level of maturity and its current climate, with a Maturity Assessment.and Safety Perception Survey. Both are excellent ways to identify areas needing improvement. Those areas provide great material for selecting objectives for your strategy, or developing tactical plans, that will help you achieve safety culture excellence.

Part III examined strategic analysis, presenting assessments for all four Strategic Areas of Safety. The insights gained from those analyses provide valuable input for strategic planning. Part IV will examine the strategic planning process itself, and how to execute strategic plans successfully.

Endnotes

1. "Achieving Safety Culture Excellence," *Health and Safety Laboratory*, accessed November 7, 2023, https://www.hsl.gov.uk/what-we-do/safety-culture#:~:text=Safety%20culture%20is%20a%20combination,how%20it%20should%20be%20done.

2. Ronny Lardner, Mark Fleming PhD, and Phil Joyner, "Toward a Mature Safety Culture, *IChemE Symposium Series*, accessed November 8, 2023, https://www.icheme.org/media/10194/xvi-paper-49.pdf.

3. Terry L. Mathis, "Safety and Performance Excellence: Behind the Bradley Curve," *EHS Today*, August 8, 2016, accessed November 2, 2023, https://www.ehstoday.com/safety/article/21918409/safety-and-performance-excellence-behind-the-bradley-curve.

4. Terry L. Mathis, "Safety Excellence Maturity Model," *EHS Today*, February 27, 2019, accessed November 2, 2023, https://www.ehstoday.com/safety-leadership/article/21920022/safety-excellence-maturity-model.

5. "Safety Culture Maturity ® Model," *The Keil Centre*, accessed November 7, 2023, https://keilcentre.co.uk/services/human-factors-ergonomics/safety-culture/scmm/.

6. Chris Goulart, "Resolving the Safety Culture/Safety Climate Debate," *Occupational Health & Safety*, accessed November 12, 2023, https://ohsonline.com/Blogs/The-OHS-Wire/2013/11/Resolving-the-Safety-CultureSafety-Climate-Debate.aspx.

7. "dss+ Bradley Curve™," *dss+*, accessed November 15, 2023, https://www.consultdss.com/transform-culture/dss-bradley-curve/.

8. Patrick Hudson, "Safety culture: The ultimate goal," *Flight Safety Australia*, September-October 2001, 30, accessed November 9, 2023, https://www.skybrary.aero/sites/default/files/bookshelf/1091.pdf.

9. "Safety Culture Maturity® Model," *The Keil Centre*, accessed November 7, 2023, https://keilcentre.co.uk/services/human-factors-ergonomics/safety-culture/scmm/.

10. Patrick Hudson, "Safety Culture and Leadership," *Safe Work Australia*, accessed November 12, 2023, https://youtu.be/4x98hwtvbBU?si=jX7mihZuYjfNTHA9.

11. Ibid.

12. Dan Petersen, *Safety Management, a Human Approach, Third Edition*, (Des Plaines: American Society of Safety Engineers, 2001), 336.

13. "Employee Safety Surveys," *nsc*, accessed November 15, 2023, https://www.nsc.org/workplace/safety-services/employee-perception-surveys.

14. "dss+ Safety Perception Survey™", *dss+*, accessed November 15, 2023, https://www.consultdss.com/dss-safety-perception-survey/

15. "CCC Safety Culture Assessment," *Culture Change Consultants*, accessed November 15, 2023, https://www.culturechange.com/solutions/safety-culture-assessment/.

16. "Safety Perception Surveys," *ProAct Safety*, accessed November 15, 2023, https://proactsafety.com/solutions/consulting/safety-perception-surveys.

17. "NSC Surveys: Frequently Asked Questions," *nsc*, accessed November 15, 2023, https://www.nsc.org/workplace/safety-services/employee-perception-surveys/nsc-surveys-frequently-asked-questions#Q5.

18. Steven I. Simon, "Measurement as a Transformative Tool: The Culture Assessment," *Culture Change Consultants*, accessed November 12, 2023, https://www.culturechange.com/wp-content/uploads/2014/10/safetycultureassessment.pdf

19. "What is a Good Employee Survey Response Rate?" *Culture Amp*, accessed November 15, 2023, https://www.cultureamp.com/blog/what-is-a-good-survey-response-rate.

20. Ibid.

21. Diago Graglia, "What is a good survey response rate?" *SurveyMonkey,* accessed March 18, 2026, https://www.surveymonkey.com/curiosity/how-many-people-do-i-need-to-take-my-survey/.

22. Ibid.

23. "dss+ Safety Perception Survey Brochure," *dss+*, accessed May 16, 2025, https://www.consultdss.com/content-hub/safety-perception-survey-brochure/

PART IV
Strategic Planning

Chapter Ten:
Laying the Foundation

A vision statement is the anchor point of any strategic plan. — Cascade

We know that a house built on a weak foundation will not stand. The same is true for commercial buildings. Both have requirements to meet for a foundation's strength. In the construction industry, the controlling contractor must certify that the concrete foundation has achieved the minimum compression strength before structural steel erection can begin. Building cannot start until the foundation is ready.

Next comes the erection of structural steel, the skeletal framework around which the building is constructed. It is engineered and designed to withstand the structural forces it will be subjected to. This is all done so that the building can fulfill its purpose for decades to come.

A solid foundation is also important for strategic planning. The foundation, as well as the framework built above it, will be subjected to forces that could make or break it. Both should be designed to withstand those forces.

Strategic planning became popular among corporations during the 1960s, and companies still use it today to help achieve their business goals and objectives. Unfortunately, it has not seen the level of popularity in the safety profession that it deserves.

What is Strategic Planning?

Strategic planning answers the questions: "Where are we, where do we want to go, and how do we get there?" It establishes the organization's goals and objectives, prioritizes its activities, allocates its resources, and aligns its stakeholders in achieving those goals and objectives. Organizations usually apply this process at the corporate or business unit levels. They can and should apply it at the functional level, as well.

The Corporate Finance Institute (CFI) is the largest finance-only training and skill development platform in the world. In their online article titled "Strategic Planning: Build a Clearer Path to Business Success," CFI answers the question, "What is Strategic Planning?" It states:

> Strategic planning is the art of creating specific business strategies, implementing them, and evaluating the results of executing the plan, in regard to a company's overall long-term goals or desires. It is a concept that focuses on integrating various departments (such as accounting and finance, marketing, and human resources) within a company to accomplish its strategic goals.[1]

The CFI definition captures three key aspects of the process worthy of consideration, it is *a process,* it is *an art,* and it promotes *integration.*

A Process

One key aspect is that strategic planning is a process, one that involves creating, executing, and evaluating strategies. Strategic planning is an ongoing process, not a one-time event. You don't just create a strategic plan, you also have to execute it, which often requires some modifications along the way.

To ensure continual improvement, the process should follow the W. Edwards Deming PDCA (Plan-Do-Check-Act) Cycle, or some other continual improvement model. After creating the strategic plan, the organization must execute it properly and evaluate it periodically. Poor execution and progress tracking are the primary reasons strategic plans do not achieve their goals and objectives.

An Art

Another key aspect of the CFI definition is the viewpoint that strategic planning is an art. Strategic planning, although systematic, analytical, and supported by verifiable data and sound reasoning, is not an exact science. Regardless of how good you think your strategic plan is, there is no guarantee it will achieve its objectives, even when properly executed.

There are two primary reasons for this. First, it relies on how well the planned actions lead to the desired outcomes. Strategic planning—determining how to get from where you are to where you want to be—hinges on that very point. A literal road map can get you to your destination, but a figurative one may not.

Robert Anthony was an organizational theorist, Harvard Business School (HBS) professor of management control, and one of the early thought leaders on strategic management. He viewed strategic planning as an art, not a science. In an HBS working paper titled "Conceptual Foundations of the Balanced Scorecard," author Robert S. Kaplan, one of the co-creators of the Balanced Scorecard approach, quoted Anthony.

> Foreshadowing the subsequent development of strategy maps, Anthony claimed that strategic planning depends "on an estimate of a cause-and-effect relationship between a course of action and a desired outcome," but concluded that, because of the difficulty of predicting such a relationship, "strategic planning is an art, not a science."[2]

An actual cause-and-effect relationship between a planned action and the desired outcome is not guaranteed, thus adding a level of uncertainty to the process.

Another reason strategic planning is an art is the same reason Sun Tzŭ titled his book *The Art of War*. Circumstances change, and even the best-laid plans need course corrections. This makes the strategy of war an art, not a science. If you recall from Chapter Two, Prussian Field Marshal Helmuth von Moltke the Elder said: "Therefore no plan of operations extends with any certainty beyond the first contact with the main hostile force."[3]

Corporate strategy experts are well aware of that point. Dr. Graham Kenny[1], CEO of KMS Education and Strategic Factors, is a recognized expert in strategy and performance measurement. Kenny wrote an article for the *Harvard Business Review* titled, "Strategic Plans Are Less Important than Strategic Planning." In his article, Kenny used that quote from Moltke, as well as related quotes from Winston Churchill and Dwight D. Eisenhower. Churchill stated: "Plans are of little importance, but planning is essential," and Eisenhower stated: "Plans are worthless, but planning is everything."[4] All three quotes convey the idea that it is the process, not the end product, that is most important in strategic planning.

Kenny then made the following comparison between military and business strategy:

> Like military strategy, business strategy is developed and applied in a fluid, unpredictable environment, and the distinction that Moltke, Churchill, and Eisenhower draw between planning and the plan is very pertinent for senior executives charged with crafting a company's strategy.[5]

Integration

A third key aspect of the CFI definition is that the strategic planning process must integrate the organization's various departments to achieve its goals and objectives. This is where functional strategic plans enter the picture.

ArchPoint Consulting is a firm specializing in strategic planning and execution. In their online article, "Fascinating Statistics on Why Corporate Strategy Breaks Down," ArchPoint explains why cascading strategic plans down through the organization is important.

> The act of cascading the strategy from the corporate level down to the functional level will provide a great amount of clarity for your teams. Seeing the activity that must happen at their level–and how that activity connects back to the overall plan– is critical in aligning the organization.[6]

There are three levels of corporate strategy, the corporate, the divisional or business unit, and the functional levels. ClearPoint Strategy markets strategic planning software that brings together all the tools needed to execute a strategic plan. In an online article titled "Breaking Down the Three Levels of Strategy in Any Business," ClearPoint' states:

> Having a solid understanding of these levels of strategy will help you break your strategy into the correct levels, so you can align your company-wide goals from the top of your organization (the corporate level) to the bottom (the functional level). Additionally, if you approach your strategy using these three levels, leaders across your organization will have a better understanding of how their strategic activities impact your company's high-level strategy.[7]

Below is a brief description of the three levels.

1. *Corporate Level:* This is the highest level of strategy for corporations, which are composed of various business lines or divisions. The corporate strategy is broader and more strategic. It addresses how the corporation will allocate its resources, optimize its performance, and ensure its growth.

2. *Divisional or Business Unit Level:* Having strategic plans at this level helps the corporate office to align its strategic plan with its various divisions or business units, and to evaluate their success. Key objectives of strategic plans at this level are differentiating itself from its competitors and gaining a competitive advantage over them.

3. *Functional Level:* This is the lowest level of strategic plans and includes individual strategies for the various departments within the corporation or business unit. Functional strategic plans need to provide specific details, requiring strategic, operational, and tactical plans.

Benefits of Strategic Planning

Strategic planning has many benefits at the functional level. Some of these include:

- *Provides Direction:* Strategic planning helps EH&S departments set the direction they need to go in to improve safety performance and achieve safety excellence.

- *Promotes Collaboration:* Strategic planning motivates the EH&S department to work with all departments and levels of the organization to achieve both organizational and functional level strategic objectives.

- *Aligns Teams:* Strategic planning helps to align the EH&S team around a common set of strategic objectives and promotes ownership and accountability for these objectives

- *Emphasizes Priorities:* Strategic planning places the emphasis on those activities that the EH&S department needs to engage in to achieve its strategic objectives.

- *Tracks Performance:* Strategic planning provides a means for tracking and measuring performance as the EH&S department works toward achieving its strategic objectives.

If you want to improve your company's safety performance, or further your journey toward safety excellence, strategic planning is the process to get you there.

Strategic Planning Framework

When performing strategic planning, it helps to have a solid foundation and well-structured framework to support and guide the process from start to finish. The process of developing a strategy begins with laying the foundation, the topic of this chapter. It continues with building the framework, which is the topic of the next chapter.

Figure 11 provides a diagram of the functional-level strategic planning framework presented in the rest of Part IV. The key elements include your mission, vision and values, and a Safety Management Framework. It then adds focus areas, the Strategic Areas of Safety. Combined, these elements represent the foundation. Strategic Objectives under each focus area represent the structural framework of your Strategic Plan.

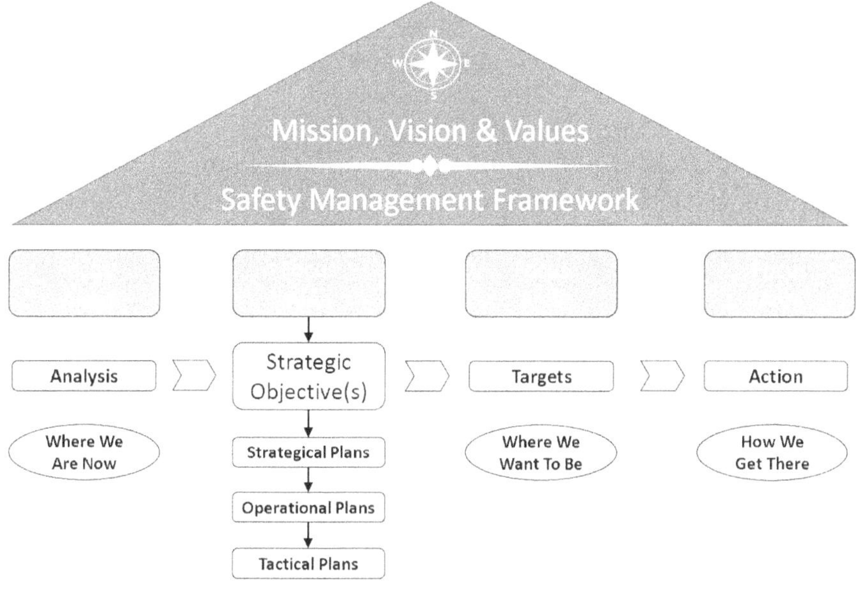

Figure 11.   Strategic Planning Framework

The strategic objectives component of this framework applies to each of the four focus areas, although figure 11 only depicts one. The process begins with analysis and ends with action. (Chapter Eleven addresses a process for selecting the right objectives.)

Functional strategic planning should be a collaborative effort between department leaders and managers, as well as business leaders. Since departments function as a team, key team members should be involved in developing the strategy to ensure the entire team agrees with it.

Because the functional strategic plan must align with and support the corporate and/or business unit strategic plan, business leaders should be involved to ensure that the various plans align. This collaboration is most effective when it occurs from the very start of the strategic planning process, which begins with developing the department's mission, vision, and values.

Mission, Vision & Values

Many companies have mission and vision statements, and a set of core values. Your corporation likely does, too. Those statements essentially define the company's purpose, where it wants to go, and how it will behave along the way. In strategic planning, they act as blinders (i.e. blinkers), much like those used in horse racing, to keep strategists focused on what matters most to the company in achieving its goals and objectives. They also help to inspire the company's key stakeholders, including employees, customers, and shareholders.

These statements and values are not as common at the functional level. Since they play such an important part in the strategic planning process, I recommend developing functional-level statements that are more specific to the department's role in the business. Just like at the corporate and business unit level, these statements will help you keep the blinders on to focus your strategic plan on what matters most to your department.

Mission Statement

A mission statement answers the question, "Why do we exist?" It describes in clear, concise, and plain language what your department does, how you do it, and why you do it. This provides a simple three-step process you can apply to developing your mission statement, describing the answers to these three questions: what, how, and why. Consider the following potential answers to those three questions:

1. *What do we do?* Safety professionals do so many things that it is hard to be concise in describing the "what." In the mission statement, you want to get to the heart of the matter. We essentially identify, evaluate, and eliminate or control hazards, reducing the risk to people, plant, and property.

That is clear and concise language, but is it plain language that can motivate all stakeholders? It also does not capture the compliance aspect of what we do. Put another way: "We prevent incidents and injuries, preclude environmental effects, and ensure regulatory compliance." That is clear, concise, and stated in plain language that all stakeholders can understand and appreciate.

2. *How do we do it?* Similarly, we safety professionals take many approaches and employ different methods in doing what we do. As much as we would like stakeholders to know every detail about how we do safety, this would be an extensive list. The mission statement is not the format for it. The mission statement should be a high-level description of how we go about reducing risk and complying with regulations.

While your department might be slightly different, many safety departments do this by developing a proactive and participative safety culture, engaging all levels of the organization in the safety process, and implementing effective and compliant management systems that identify and eliminate or control hazards.

3. *Why do we do it?* We do safety to protect people, plant, and property, to state it in a nutshell, right? It is a common phrase used by safety professionals who fully understand why we do what we do. That phrase, however, does not convey the fact that we truly care about people and the environment. Conveying that fact would inspire more support from stakeholders.

An inspirational declaration would be: "We do this because we care about our people and want them to go home safe every day, and because we care about keeping our environment as clean as possible."

Consider the following example safety function mission statement. Is it written in clear, concise, and plain language? Will it inspire others to support the department's efforts? You be the judge.

Our Mission

We are committed to reducing injuries, preventing environmental impacts, and ensuring regulatory compliance. We do this by creating a culture of caring, developing employees who are competent in safety, and by implementing management systems that identify and eliminate or control hazards. We do this because we care about our people and want them to go home safe every day. We also care about keeping the environment clean for future generations.

Vision Statement

According to Cascade:

> A vision statement is the anchor point of any strategic plan. It outlines what an organization would like to ultimately achieve and gives purpose to the existence of the organization. A good vision statement should be short, simple, specific to your business, and leave nothing open to interpretation. It should also have some ambition.[8]

As a safety department, your vision statement should describe the future state of safety that your team aims to achieve. It should be aspirational, requiring the team to stretch their knowledge and skills to achieve it, yet attainable, so that it is not demotivating.

How does the vision statement discussed here relate to the strategic vision discussed in Chapter Five? Essentially, they are one and the same. In Chapter Five, a strategic vision was created for a more narrowly focused strategy. Here, a broader vision is needed, but it is still a strategic vision.

Vision statements can be written from three different perspectives. These include:

1. What you want to achieve regarding safety.

2. What you want to have as a safety program or culture.

3. What you want to be as a department or business where safety is concerned.

Below are some examples of statements from different perspectives.

Our Vision

- Achieve safety excellence with a caring culture that strives for zero injuries, and environmental sustainability driven by a commitment to stewardship.

- Have a culture where everyone is committed to and engaged in creating a work environment where no one is injured, while maintaining a natural environment that is both healthy and sustainable.

- Be a department that is the driving force behind instilling the value of safe and environmentally conscious behavior throughout the company.

- Be a company that integrates environmental, health, and safety excellence into all aspects of the business, and maintains the highest level of compliance.

Some companies include a list of descriptions in their vision statement, which, if kept short and simple, can have the same impact and motivation as a single statement. Whatever perspective or form your vision statement takes, it needs to align with your company's vision, but it also needs to be as meaningful as possible to you and your department. It needs to be your vision, and it needs to motivate you and your team to achieve it.

Core Values

Values are an important aspect of an organization's culture. They shape the organization's norms, its standards of behavior. A company's core values are principles intended to guide how employees behave while doing business, and influence their decision-making process.

The corporate or business unit core values need to be considered when developing a functional strategic plan, but you might also find it beneficial to augment those core values with some that are specific to your function. Those function-specific values would further guide how your department goes about fulfilling its role within the organization.

Below are example values that help guide the safety function.

- *Professionalism:* We provide our services in a professional and ethical manner, offering sound safety advice with viable strategies and solutions.

- *Innovation:* We deliver value-based solutions by introducing new ideas, methods, or plans that have a positive impact on safety and the business overall.

- *Collaboration:* We collaborate with all stakeholders in achieving common environmental, health and safety goals, and value all perspectives in doing so.

- *Caring:* We care about the well-being of every employee, and help employees to care equally about their co-workers.

When crafting function-specific values, it is important to engage key members of the department and find out what values are important to them. These values should reflect the values of those performing the function.

Safety Management Framework

A Safety Management Framework is a set of essential elements of an overall safety program. This book provides an example framework to introduce the concept, and describe how it relates to a safety strategy.

The first strategic benefit to Safety Management Frameworks is that they depict everything that managing an effective safety program encompasses.

W. Edwards Deming was an American statistician, educator, and consultant who helped Japanese businesses recover after WWII through quality control methods. He stated: "If you can't describe what you are doing as a process, you don't know what you're doing."[9] Having a Safety Management Framework tells your stakeholders that you know what you are doing when it comes to managing safety.

Figure 12 provides a diagram of a framework based on a heavy industry application. The core of the diagram includes the department's safety vision and values, which guide all safety activities. The inner wheel represents the three major categories or phases of the process. They include: *foundation*, fundamental management elements; *execution*, elements related to safe work execution; and *examination*, elements related to assessing and improving overall performance. The outer wheel contains twelve key elements, rotating clockwise starting at the top with visible leadership, representing a continual improvement process.

When creating a Safety Management Framework, you should clearly define what each of the elements incorporate. You can do this by listing a set of sub-elements under each element. Including three to five sub-elements should suffice. The sub-elements in my framework were shown previously. In Chapters Eight, table 24 listed the sub-elements for the foundation elements. In Chapter Nine, table 27 included sub-elements for the execution elements, and table 28 included sub-elements for the examination elements.

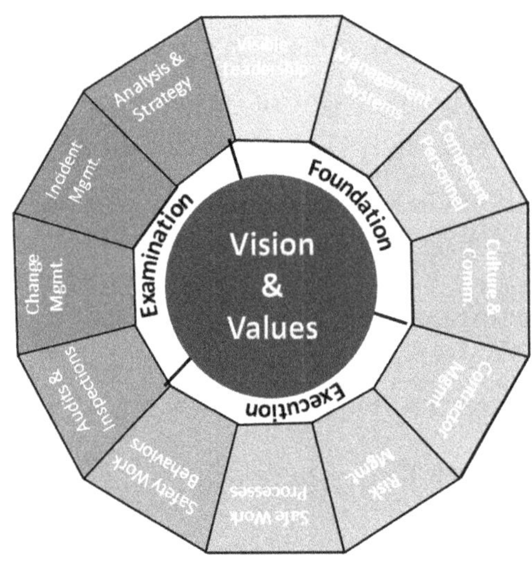

Figure 12.   Safety Management Framework

The second strategic benefit of a Safety Management Framework is that it provides you with an assessment tool. The framework elements can help you assess competency levels of safety professionals and other key positions within the organization, as shown in Chapter Eight. The elements can also help you assess the maturity level of your overall safety program, and your culture, as shown in Chapter Nine.

Between the elements and sub-elements, the Safety Management Framework goes a long way in describing what you do and how you do it. I took it one step further and created a document that explained the reasons why we do what we do. It provided all the details behind the high level mission statement, for those parts of the organization who needed to know those details.

Focus Areas

Chapter Three presented the four Strategic Areas of Safety as the foundation stones of a safety strategy. They make ideal focus areas for strategic analysis and planning purposes. Developing safety strategies under those focus areas will help you avoid majoring in the minors. It will also help to ensure that you are allocating resources in the most effective way for achieving your safety goals and objectives.

Another aspect of focus areas is that they can provide you with key measures of success. As with the Balanced Scorecard approach, strategic safety uses measures to determine the level of success in achieving safety goals and objectives. Each of the Strategic Areas of Safety has the potential for developing overarching performance measures.

Laying a foundation for your strategic planning is a very beneficial exercise. Even if you decide not to develop a functional mission statement and core values, I believe a functional vision statement is too important of a step to skip. It is your strategy's anchor point. Designing a framework for your strategic plan, namely setting objectives under each focus area, is also critically important, and doing without them would leave you without a true strategy.

As with building construction, laying a proper foundation for your strategic plan is an integral step in the planning process. A solid foundation and properly designed framework will ensure your strategy supports your mission and achieves your vision. Once you have laid the foundation, you can start building your strategic plan, which is the topic of the next chapter

Endnotes

1. "Strategic Planning," *Corporate Financial Institute*, accessed February 20, 2022, https://corporatefinanceinstitute.com/resources/knowledge/strategy/strategic-planning/.

2. Robert S. Kaplan, "Conceptual Foundations of the Balanced Scorecard," *Harvard Business School*, 2010, 6, accessed February 20, 2022, https://www.hbs.edu/ris/Publication%20Files/10-074_0bf3c151-f82b-4592-b885-cdde7f5d97a6.pdf.

3. Helmuth von Moltke, "Plan of Operations" (1871-81)," *Moltke on the Art of War: Selected Writings*, ed. Daniel J. Hughes (New York: Ballantine Books, 1993), 92.

4. Graham Kenny, "Strategic Plans are Less Important than Strategic Planning." *Harvard Business Review*, June 21, 2016, accessed February 20, 2022, https://hbr.org/2016/06/strategic-plans-are-less-important-than-strategic-planning.

5. Ibid.

6. "Fascinating Statistics on Why Corporate Strategy Breaks Down," *ArchPoint Consulting*, accessed February 20, 2022, https://archpointconsulting.com/strategy/fascinating-statistics-on-why-corporate-strategy-breaks-down#.

7. Jenna Weaver, "Breaking Down the Three Levels of Strategy in Any Business," *ClearPoint Strategy*, accessed February 21, 2022, https://www.clearpointstrategy.com/levels-of-strategy/.

8. "How to Write a Strategic Plan: The Cascade Model," *Cascade*, 19, accessed February 22, 2022, https://www.cascade.app/how-to-write-a-strategic-plan.

9. W. Edwards Deming, "Science Quotes by W. Edwards Deming," *Today in Science*, accessed November 3, 2023, https://todayinsci.com/D/Deming_WEdwards/DemingWEdwards-Quotations.htm.

Chapter Eleven:
Mapping a Strategy

One of the more powerful elements in the Balanced Scorecard framework, a strategy map, clearly visualizes your strategy for everyone to see. — Joseph Lucco

Long before electronic road navigation systems such as Garmin (1998) and Google Maps (2005), the Automobile Association of America (AAA) had the TripTik®(1937). AAA TripTiks® are personalized, hard copy, spiral-bound, highlighted strip maps that show you the route from your point of origin to your destination. They also provide driving directions, identify gas stations, list AAA approved lodging and restaurants, and points of interest along the route.

Wouldn't it be nice to have a detailed route mapped out for you that showed you how to get from where your safety program is today, to where you want it to be in the future, one that shows every step of your journey? That is essentially the end product of strategic planning. Developing a road map to get you from here to there.

Laying a solid foundation was the intention of the top half of the Strategic Planning Framework presented in Chapter Ten. The intention of the bottom half is providing the structure for developing strategic plans. This process begins with selecting the right objectives to include in your plan.

Strategic Objectives

Having already referred to the focus areas as foundation stones or cornerstones of a strategic plan, it helps to think of strategic objectives as stones, too. Think of objectives as the milestones or stepping stones that help you fulfill your mission and realize your vision. They represent what you are trying to achieve on your strategic journey.

Yigit Durdag is the former Head of Content for GrowForce, a Belgium-based consulting firm specializing in marketing growth strategies. In an online article titled "How to Write Strategic Objectives: A Detailed Guide with Examples," Durdag described strategic objectives. He wrote: "Strategic objectives are manifestations of your vision that can help you achieve your future business goals."[1] Durdag provided an excellent way to consider your strategic objectives, manifestations or images of what realizing your vision will look like in each of the Strategic Areas of Safety.

As with focus areas, a common question regarding strategic objectives is, "How many objectives should we have?" Ted Jackson is the founder and managing partner at ClearPoint Strategy. In an online article titled "How to Create & Write Out Your Strategic Objectives," Jackson wrote: "You should have no more than 15 objectives in your strategy."[2] Since ClearPoint Strategy uses the Balanced Scorecard approach to strategic management, Jackson recommends creating strategic objectives in all four of the Balanced Scorecard perspectives. Likewise, safety professionals should create strategic objectives in all four Strategic Areas of Safety.

Following Jackson's recommendation, you could have three to four objectives in each of the four Strategic Areas of Safety, if you spread these across the four areas, and remain within the recommended number.

In his article, Durdag stated: "In our opinion, you can create a minimum of two and a maximum of five strategic objectives per each of your focus areas / departments."[3] The actual number of strategic objectives that you include will be based on your unique circumstances and available resources. The number of resources at your disposal to execute your strategy is the key determining factor in how many objectives to include. You or your department might not have the resources needed to manage fifteen strategic objectives.

The real challenge in developing strategic objectives is not determining the number of objectives to include, but determining the right objectives to include. Safety professionals who take a shotgun or brute force approach to annual safety planning, although well-meaning, typically place more value on the quantity of objectives rather than their quality. This is the defining difference between a non-strategic approach to safety planning and a strategic one—choosing the right objectives.

Michael Porter is an American academic and award-winning author of competitive business strategies. In a *Harvard Business Review* article titled "What is Strategy?" Porter made one of the most often quoted statements about strategy. He wrote: "The essence of strategy is choosing what not to do."[4]

Porter was discussing trade-offs made when competing in business. If the ultimate purpose of a business strategy is gaining a competitive advantage, then it stands to reason one would only include strategic objectives that will fulfill that purpose. In safety, we too need to make trade-offs when competing for company resources, or when allocating safety department resources.

So, then, what do we want to include in our safety strategies? We should include only those objectives that are important to fulfilling our mission and realizing our vision—nothing more, nothing less.

Before we examine selecting strategic objectives, it would help to know what one should look like. Jackson stated: "The typical format of a strategic objective is 'Verb + Adjective + Noun.' If you use this formula, your strategic objectives will create an action statement. Note that your strategic objectives should describe *your* strategy—not just a typical strategy."[5]

Jackson also recommends stating objectives in two to three sentences to clarify the intent and meaning of your objectives. Some organizations also incorporate a brief bullet list of strategic plans as part of the stated objective.

Using one of the vision statement examples from the previous chapter, below is an example of a strategic objective that meets the criteria of that approach, and also incorporates a list of strategic plans.

Example Objective

Vision: We envision having a culture where everyone is committed to and engaged in creating a work environment where no one is injured, and maintaining a natural environment that is healthy and sustainable.

Strategic Area: Safety Culture

Strategic Objective: We will develop and maintain [verbs] a positive, proactive, and participative [adjectives] safety culture [noun] where all employees are engaged in the safety process [intent]. To achieve this, we will [strategic plans]

- Implement a periodic Safety Perception Survey to assess the current state of the safety culture, and measure improvement.

- Assess the maturity of our safety program and develop programs to get us to the next level of maturity.

- Develop culture improvement projects involving employees at all organizational levels to foster a culture of mutual trust and caring.

How you go about implementing your strategic plans is through the use of operational and tactical plans, similar to how the military achieves its objectives with those levels of warfare.

Strategic Objective Selection

There are various methods available to help you generate a list of strategic objectives. We will discuss one of the most popular, the S.W.O.T. Analysis. However, that method only helps you generate a list; it does not help you select the right objectives from that list. In the many online articles I have read describing a S.W.O.T. Analysis, I could not find one that explained how to select the best objectives from your generated list. Instead, they simply instruct you to choose which objectives to use, which is more subjective. I suggest using a more objective approach to selecting objectives.

In strategic safety, we want to ensure victory with our strategy, so we need to select the best possible objectives. I recommend using a method for identifying, analyzing, and selecting objectives that is comprehensive, systematic, and objective. This chapter presents such a plan. Once you learn it, you will see the benefits of adding a greater level of objectivity to your selection process. It will make selecting the best objectives for your strategy much easier.

Figure 13 illustrates a seven-step selection process. It starts with collecting and assessing the information needed to understand the current state and identify opportunities. A two-part process follows to identify potential objectives and pare them down those that are practicable and worth considering. This list is then analyzed using both a risk-based and value-based approach to select and prioritize objectives to include in your strategic plan. This process helps you to select the best possible objectives and also provides you with the supporting information to help sell your strategic plan to executive leadership.

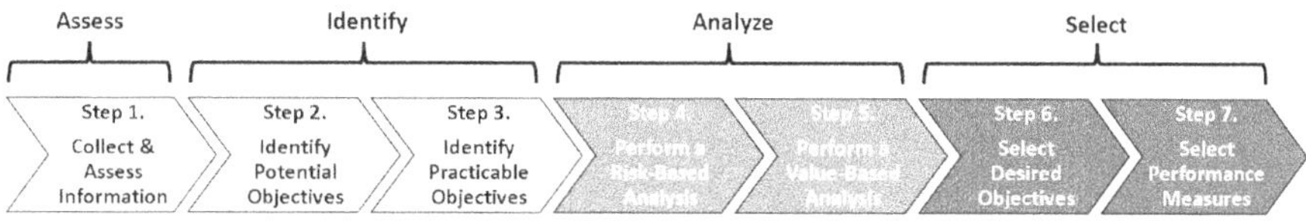

Figure 13.   Objective Selection Process

The steps of this process are described below. As you complete this process, you will want to ensure that each of your focus areas has objectives, and that potential objectives are directly associated with one of your focus areas. Any objective can apply to more than one focus area, but each will apply more to one of those areas than the others.

You could perform the process for each focus area at a time, but this could inhibit your creative and intuitive thought. It is best to just note which focus area each potential objective falls under.

Step 1 - Collect & Assess Information

Every journey has a starting point. You need to know where you are starting from before you can map out a route to where you want to go. The first step involves assessing existing information that helps you better understand where you are starting from. These assessments provide the raw material to use in step two of the process. Part III of this book devoted an entire chapter to the assessments for each of the Strategic Areas of Safety.

Step 2 - Identify Potential Objectives

This is the step of the process where a list of potential strategic objectives takes shape. It generates a list of issues or opportunities that potential objectives will be based on. Your focus should be on generating the list, not on identifying or evaluating the solutions. You will draft objective statements and select objectives later in the process. For now, you are in brainstorming mode and are after quantity, not quality.

The most popular method used to generate this list is the S.W.O.T. Analysis. S.W.O.T. is an acronym for: *strengths, weaknesses, opportunities,* and *threats,* four keywords associated with questions regarding your organization. This method allows you to assess internal and external factors, as well as your current state and future potential.

A similar version uses the acronym, P.A.C.E., which stands for: *preserve, attain, consider, and eliminate.* Performing the analysis is the same for both versions. Ask the questions and list any relevant issues under each category.

Table 33 provides the questions asked under each keyword for both the S.W.O.T. and P.A.C.E. versions of the analysis. The assessments described in Part III can provide you with the answers to some of these questions. The other option is to rely solely on the collective knowledge of the strategic planning team.

Table 33.   SWOT & PACE Analysis

SWOT Analysis

What are your strengths, things you do well that can be leveraged to improve safety?
What are your weaknesses, things you do poorly that need to be improved?
What are your opportunities, things you are not doing that would improve safety?
What are your threats, things that could impact safety but are not being addressed?

PACE Analysis

PRESERVE
What do you have that is working and want to keep or possibly improve?
ATTAIN
What do you want to achieve that you are not currently achieving?
CONSIDER
What should you consider having that you do not already have?
ELIMINATE
What do you have that adds little value and needs to be removed?

You can use either analysis, or both, to help you identify issues to base potential objectives on. A diverse group of representatives should perform this analysis. Team members must be free to suggest issues without fear of criticism, such as during a brainstorming session. Generating a complete list of ideas for further evaluation is essential during brainstorming. Judging each item's value at this stage hinders the creative process.

Table 34 shows an example of items identified by a S.W.O.T. Analysis. Note how it assigned a unique key code to each potential objective using the first letter of the associated S.W.O.T. keyword and a sequential number. This helps to identify potential objectives throughout the remaining selection steps. The table also identifies the source analysis and focus area.

You will notice that some of these items came from the analyses performed in Part III, while others did not. Items from this table and the next will be carried through the strategic objective selection process.

Table 34.   S.W.O.T. Analysis

Key	Issue Description	Source Analysis	Strategic Area
S1	Executive leadership is supportive of safety.	Collective Knowledge	Culture
S2	Management is knowledgeable in safety processes.	Competency Assessment	Competence
W1	Safety leadership is lacking in the field and no training exists.	Collective Knowledge	Competence
W2	Limited access to safety management software.	Collective Knowledge	Compliance
W3	Employee Safety Training Profile not updated in past 5 years.	Training Assessment	Competence
W4	Limited knowledge in ergonomics displayed among employees.	Injury Attribute Analysis	Competence
O1	The company is not taking advantage of new safety technologies.	Injury Attribute Analysis	Performance
O2	27% of employees feel left out of job safety planning.	Safety Perception Survey	Culture
O3	Standardize safety plans, programs, and procedures to be more professional looking and compliant.	Collective Knowledge	Competence
T1	Changes in OSHA regulations not adequately monitored.	Collective Knowledge	Compliance
T2	Hand injuries with knives and hammers are leading injury types.	Injury Attribute Analysis	Performance
T3	Musculoskeletal injuries are second leading injury type.	Injury Attribute Analysis	Performance
T4	Housekeeping is leading cause of Near Misses and top observation.	Injury Attribute Analysis	Performance
T5	Industry growth drawing experienced EH&S staff away.	Collective Knowledge	Competence
T6	Not meeting field customers' expectations regarding safety.	Customer Survey	Compliance

A P.A.C.E. Analysis helps add to the list of items or issues to evaluate using similar but different questions Table 35 is an example of items identified by a P.A.C.E. Analysis.

Table 35.   P.A.C.E. Analysis

Key	Issue Description	Source Analysis	Strategic Area
P1	Safety Suggestion Program participation continues to improve.	Observations & Suggestions Review	Culture
P2	Compliance audits are performed, but much improvement is needed.	Audit Attribute Analysis	Compliance
A1	Improve safety training completion to ninety (90) percent on-time.	Training Assessment	Compliance
A2	Improve number of safety certifications among EH&S staff.	Competency Assessment	Competence
A3	Implementation of Safety Awareness Campaigns.	Collective Knowledge	Culture
A4	Zero OSHA citations. (Received 3 in last 5 years.)	Collective Knowledge	Compliance
C1	Purchasing Safety Management System software.	Collective Knowledge	Competence
C2	Computer-Based Training system for general employee safety training.	Training Assessment	Compliance
E1	Remove EH&S Manager Daily Inspection Reports.	Collective Knowledge	Compliance

As previously stated, online articles about strategic planning that promote the use of a S.W.O.T. Analysis stop at generating the list. The question then becomes, "How do you decide which ones to include in your strategic plan?" Decisions about strategy need to be made, for lack of a better term, strategically.

Step 3 - Identify Practicable Objectives

In this step you will analyze the list of issues generated during the S.W.O.T. and P.A.C.E. Analysis and narrow it down to those that are worth considering for your strategic plan. An excellent method for performing this is called a P.I.C.K. Analysis. P.I.C.K. is an acronym for: *possible, implement, challenge,* and *kill,* which represent the four quadrants in a P.I.C.K. Matrix. The process involves placing each potential into one of the quadrants.

The analysis uses two ratings to place each item in the matrix, an *impact* (Y axis) rating and an *effort* (X axis) rating. Each item is rated for its impact on safety, using a scale of one to ten, with one being minimum impact and ten being maximum impact. The item is then rated, using the same one to ten scale, on the amount of effort needed to address the issue or implement a related objective. These ratings provide the points on the matrix that determine which quadrant an item falls in.

Figure 14 shows the P.I.C.K. Matrix quadrants and their descriptions to help you pick which items to consider further, narrowing down your list to only those potential objectives that are most impactful and practicable.

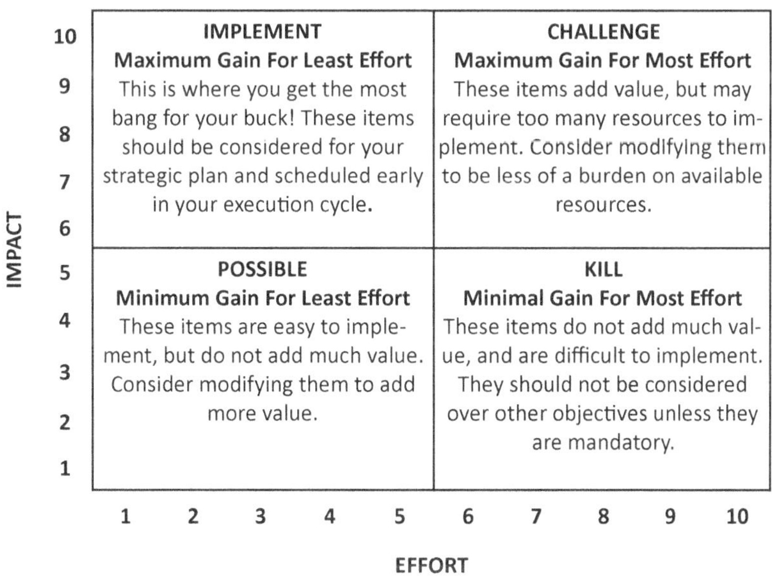

Figure 14.   P.I.C.K. Matrix

Following the P.I.C.K. Analysis, a table should be created showing both items excluded from and included for further consideration.

Table 36 shows the list of items that were included for further evaluation, their draft objective statements, and P.I.C.K. Matrix quadrant. (Excluded items have been filtered out, and P.I.C.K. impact and effort ratings are not displayed.)

Table 36.   P.I.C.K. Analysis (Items Included)

Strategic Key	Area	Issue Description	Draft Objective	P.I.C.K. Matrix
W1	Competence	Safety leadership is lacking in the field and no training exists.	Develop safety leaders within the organization.	Implement
O2	Culture	27% of employees feel left out of job safety planning.	Create a caring and collaborative safety culture.	Implement
T2	Performance	Hand injuries with knives and hammers are leading injury types.	Create a workplace and mindset that keeps hands out of the line of fire.	Challenge
T3	Performance	Musculoskeletal injuries are second leading injury type.	Significantly reduce musculoskeletal injuries.	Implement
T6	Compliance	Not meeting field customers' expectations regarding safety.	Exceed all client EH&S expectations.	Implement
P2	Compliance	Compliance audits are performed, but much improvement is needed.	Improve the Compliance Audit Program.	Implement
A1	Compliance	Improve safety training completion to ninety (90) percent on-time.	Improve the Compliance Training Program.	Implement
A2	Competence	Improve number of safety certifications among EH&S staff.	Implement an EH&S Professional Development Program.	Implement
A3	Culture	Implementation of Safety Awareness Campaigns.	Implement quarterly safety awareness campaigns.	Implement
A4	Compliance	Zero OSHA citations. (Received 3 in last 5 years.)	Eliminate OSHA citations and fines.	Challenge

Adding notes during the P.I.C.K. Analysis explaining your choices may be useful in subsequent strategic planning efforts. Perhaps an item or related objective would do little toward fulfilling your mission or realizing your vision. The item might already be addressed in your existing programs. Maybe you combined two similar items, or treated one as a strategic plan under another item selected as an objective.

As you can see, this is the time to draft objective statements for the items you select for further consideration. The next two steps of the process are better served by having draft objective statements as opposed to issue descriptions. These steps include analyses that will evaluate your potential objectives from a risk-based and value-based perspective.

The analyses described below are based on those presented by David Downs, President of EHS Management Partners, and William Heim, Principal Consultant of Alliance Health & Safety, during a seminar at the then American Society of Safety Engineers' 2014 SeminarFest. They referred to it as the "Business Objective-Weighted Risk Assessment."[6] These analyses provide a significant benefit to strategic planning —adding objectivity to the selection process.

Step 4 - Perform a Risk-Based Analysis

If the essence of safety is mitigating risk, you want to select objectives that address the highest risks. Therefore, it is advisable to perform a risk-based analysis of each of your potential objectives. A 5 x 5 Risk Matrix is well-suited for this purpose. For each objective, consider the EH&S risk(s) associated with the underlying issue if you *did not* pursue the objective. For example, if your objective is to develop an EH&S compliance audit program, what would the risk to safety be if you did not audit compliance?

Using the common EH&S risk ratings for both Likelihood (Frequency) and Severity (Impact), you would generate a Risk Score for each objective.

(Eq. 6)   *Risk Score = Likelihood Rating X Severity Rating*

The higher the risk score, the greater the risk, and thus the more impactful your potential objective would likely be at reducing risk. You can adopt the basic safety 5 x 5 Risk Matrix shown in figure 15 to perform the risk analysis. A score of ten or higher is worthy of consideration.

To reiterate, you are rating the potential risk to safety if the objective is *not* selected, since you would not be mitigating the associated risk. This exercise provides a relative estimate of the potential risk for not selecting a potetial objective, or, in other words, the extimated risk reduction by selecting it.

5 x 5 Risk Matrix			Severity (Impact)				
			1	2	3	4	5
			First Aid Case, Minimal Impact	Medical Treatment Injury, Minor Impact	Injury w/Temp. Disability, Serious Impact	Permanent Injury, Significant Impact	Fatality or Multiple Injuries, Major Impact
Likelihood (Probability)	5	Very Likely, Daily to Monthly	5	10	15	20	25
	4	Likely, Many Times Yearly	4	8	12	16	20
	3	Probable, Once a Year	3	6	9	12	15
	2	Unlikely, Every 5 Years	2	4	6	8	10
	1	Very Unlikely, Once in a Lifetime	1	2	3	4	5

Figure 15.   Risk Matrix

Table 37 shows the results of a risk-based analysis performed using the items listed in table 36.

Table 37.   Risk-Based Analysis

Key	Objective Statement	Likelihood / Frequency	Severity / Impact	Risk Score
W1	Develop safety leaders within the organization.	3	3	9
O2	Create a caring and collaborative safety culture.	4	3	12
T2	Create a workplace and mindset that keeps hands out of the line of fire.	4	3	12
T3	Significantly reduce musculoskeletal injuries.	3	4	12
T6	Exceed all client EH&S expectations.	4	3	12
P2	Improve the Compliance Audit Program.	3	3	9
A1	Improve the Compliance Training Program.	3	2	6
A2	Implement an EH&S Professional Development Program.	2	2	4
A3	Implement quarterly safety awareness campaigns.	4	3	12
A4	Eliminate OSHA citations and fines.	3	3	9

Step 5 - Perform a Value-Based Analysis

Since your strategic plan must align with your corporate or business level strategic plan, including the company's mission, vision, and values, it helps to determine how well your potential objectives align with the objectives of the organization's strategic plan, or how well they align with its core values.

A two-step process for performing this analysis will be presented here. The first step is to rank or prioritize the organization's strategic objectives, or its core values, whichever you prefer to use. A paired comparison is used to perform this ranking. Figure 16 provides an example of a Paired-Comparison Matrix for core values (CV) of an organization.

Paired-Comparison	Safety	Service	Integrity	Innovation	Quality	
Safety vs >		5	5	5	5	
Service vs >	1		1	5	5	
Integrity vs >	1	5		5	5	
Innovation vs >	1	1	1		1	
Quality vs >	1	1	1	5		

Figure 16.   Paired-Comparison Matrix

To perform the comparison:

1. Enter the organizational level strategic objectives or core values (CV) in the column and row headers, as shown, in the same order.

2. Compare the first item (CV1), to the other items in that row based on the column heading. If it ranks higher than another item, enter five. If it ranks lower, enter one. (In the example, Safety is rated higher than all other core values; thus, five is entered in each column.)

3. Enter the opposite value (one or five) in the corresponding reverse comparison square (i.e. Since the Safety vs. Service score is five, the Service vs. Safety score must be one, and so on.)

4. Complete the same process for the remaining rows of items.

5. Add the values in each row and place the sum in the Paired-Comparison (P-C) Score column.

In a paired-comparison, as in an FMEA, a higher P-C Score means a higher ranking. As seen in the matrix, Safety (CV1) ranked highest in importance, followed by Integrity (CV3). This can be a subjective ranking process, so it is helpful to have the entire strategic planning team involved.

The second step of this value-based analysis rates your potential objectives against the organization's strategic objectives or core values. Table 37 provides a weighted value assessment of items from table 36.

Table 38.   Weighted Value Assessment

Key	Objective Statement	Safety	Service	Integrity	Innovation	Quality	Value Score
		20	12	16	4	8	
W1	Develop safety leaders within the organization.	4	4	2	2	1	176
O2	Create a caring and collaborative safety culture.	4	3	2	2	1	164
T2	Create a workplace and mindset that keeps hands out of the line of fire.	5	2	3	4	1	196
T3	Significantly reduce musculoskeletal injuries.	5	2	3	2	1	188
T6	Exceed all client EH&S expectations.	3	4	2	1	1	152
P2	Improve the Compliance Audit Program.	4	4	4	2	1	208
A1	Improve the Compliance Training Program.	4	4	2	2	1	176
A2	Implement an EH&S Professional Development Program.	4	3	3	1	1	176
A3	Implement quarterly safety awareness campaigns.	4	2	2	1	1	148
A4	Eliminate OSHA citations and fines.	5	4	4	1	1	224

To complete the analysis, rate each potential objective based on how well it would affect achieving the organization's strategic objectives, or the strength of correlation it has with the organization's core values. The rating scale used is one to five, as described below.

1=Little or No Impact or Correlation

2=Minor Impact or Correlation

3=Moderate Impact or Correlation

4=Significant Impact or Correlation

5=Major Impact or Correlation

Calculation of the Value Score uses the associated ratings and P-C scores, as shown below, with potential objective T2 in table 37 serving as an example. The higher the value score, the more impactful the objective will be in achieving the organization's strategic objectives, or the stronger it correlates to the organization's core values. The equation below shows the value score equation.

(Eq. 7) $Value\ Score = (Rating \times P\text{-}C\ Score)_1 + (Rating \times P\text{-}C\ Score)_2 + (Rating \times P\text{-}C\ Score)_3 + (Rating \times P\text{-}C\ Score)_4 + \ldots + (Rating \times P\text{-}C\ Score)_n$

(Eq. 8) *Potential Objective 1 Value Score = (5 x 20) + (2 x 12) + (3 x 16) + (4 x 4) + (1 x 8) = 196*

(Eq. 9) *Potential Objective 1 Value Score = 100 + 24 + 48 + 16 + 8 = 196*

Having performed both a risk-based and value-based analysis, you now have two objective data points for selecting which objectives to include in your strategic plan. However, one objective could have the highest risk score, while another has the highest value score. Which should you select?

Downs and Heim went one step further and multiplied the risk score by the value score to arrive at a combined Risk-Value Score to help prioritize your potential objectives. That score is like the Risk Priority Number (RPN) in an FMEA. The highest Risk-Value Score has the highest impact.

Table 39 combines the Risk and Value Scores from tables 36 and 37, multiplies them together, and then ranks them in order of priority based on highest to lowest R-V Score.

Table 39.   Risk-Value Rankings

T2	Performance	Create a workplace and mindset that keeps hands out of the line of fire.	12	196	2352	1
T3	Performance	Significantly reduce musculoskeletal injuries.	12	188	2256	2
A4	Compliance	Eliminate OSHA citations and fines.	9	224	2016	3
O2	Culture	Create a caring and collaborative safety culture.	12	164	1968	4
P2	Compliance	Improve the Compliance Audit Program.	9	208	1872	5
T6	Compliance	Exceed all client EH&S expectations.	12	152	1824	6
A3	Culture	Implement quarterly safety awareness campaigns.	12	148	1776	7
W1	Competence	Develop safety leaders within the organization.	9	176	1584	8
A1	Compliance	Improve the Compliance Training Program.	6	176	1056	9
A2	Competence	Implement an EH&S Professional Development Program.(DO NOT INCLUDE)	4	176	704	10

Prioritizing your strategic objectives plays a critical role in helping you not only select the most impactful objectives, but also allocate resources for the greates inpact.

For financial resources, some strategic plans may involve capital expenditures (CapEx) requiring a Cost-Benefit Analysis (CBA). CapEx represents funds used by a company to undertake new projects or investments. In order to get a CapEx approved, you often need to show a return on investment (ROI) greater than the discount rate your company uses.

A CBA is the process used to measure the financial benefits of investments by comparing the increased revenues or cost-savings of a proposed project to its initial and recurring costs to determine the project's ROI. If the ROI exceeds the discount rate, the investment makes financial sense.

How to perform a CBA is beyond the scope of this book. If you are not familiar with the process, consult with your financial department, since they likely have an existing process and can help you complete it.

Step 6 - Select Desired Objectives

The first five steps of this process provided the best objectives for you to choose from. The selection process works best when performed by a diverse group of representatives from within the organization. Having the strategic planning team involved in selecting the final list of objectives is critically important, since the strategic plan will need the support of the entire organization to be executed. The best way to get support is by involving them in the decision-making process. You also need their help in communicating the strategy to other departments, selling them on the benefits of it, as well as gaining their support in executing the strategy.

I trust you now see the benefit of using a comprehensive, systematic, and objective process for selecting your strategic objectives. It helps you to select the most impactful objectives, prioritize them for allocating resources, and sell them to executive leadership and the rest of the organization.

Step 7 - Select Performance Measures

One more question to ask regarding strategic objectives is, "How will you know when you achieve them?" A strategic objective must include measures of success to track progress and effectiveness.

Measures are a central concept in the Balanced Scorecard approach; so much so that *scorecard* is part of its name. After selecting your objectives, you must select their measures. Robert S. Kaplan, in another *Harvard Business Review* article titled, "Conceptual Foundations of the Balanced Scorecard," wrote the following regarding objectives and performance measures, and the benefit of starting with your objectives:

> Thus, while our initial article had a subtitle, "Measures that Drive Performance," we soon learned that we had to start not with measures but with descriptions of what the company wanted to accomplish. It turned out that selection of measures was much simpler after company executives described their strategies through the multiple strategic objectives in the four BSC perspectives.[7]

There are two primary types of performance measures to consider when developing your strategic safety plan; Objectives and Key Results (OKRs), and Key Performance Indicators (KPIs). While similar, they serve different purposes. OKRs define what an organization wants to achieve and how it will define achieving it. The first part is the objectives, which have already been determined in this selection process. The second part is the key results, the metric(s) that define what achievement looks like. They are lagging indicators and are always associated with an objective.

KPIs are metrics used to track the progress and completion of action plans designed to achieve the organization's objectives. They measure how well you are performing against your plans, not your objectives. Because they measure the process and not the results, they are leading indicators.

Recall from Chapter Ten that strategic planning is an art, not a science. This is because the presumed cause-and-effect relationship between a planned action and desired results adds a level of uncertainty to the strategic planning equation. For this reason, both OKRs and KPIs are needed for measuring a strategic safety plan. They measure the two critical aspects of plan performance and confirm any correlation between actions and results.

First, KPIs measure how well the organization is performing at executing the strategic, operational, and tactical plans. Second, the key results portion of OKRs measures how well the strategic plans are performing at achieving the strategic objectives. Theoretically, your organization can have a flawless performance in executing the plan without the plan achieving the objectives. You need both measures to determine where the fault lies if you do not achieve your objectives, poor execution or little cause and effect. Consider the following example:

Example Objectives, OKRs & KPIs

Strategic Objective: Significantly reduce musculoskeletal injuries.

Key Result: Year-on-Year Reduction in Recordable Musculoskeletal Injury Rate of 25% Each Year.

That key result, a lagging indicator, would reflect achievement of the objective. It not only measures achievement, but it also tracks progress toward achieving the goal, such as monitoring the reduction quarterly, throughout your strategy execution period.

However, if you want to ensure that the strategy is being executed according to plan, you also need to monitor the progress of the associated strategic plans. That is where KPIs come into play. Below are examples of KPIs used to track progress associated with this particular strategy execution:

- 95% of employees trained in ergonomic hazards by Q1 of year one.

- 100% completion of ergonomic assessments (RULA, REBA, etc.) of strenuous and highly repetitive jobs by Q1 of year one.

- Ergonomic improvement plans in place by Q2 of year one for all jobs flagged as needing them by the ergonomic assessments.

- 75% completion of ergonomic improvements in year one, with the remaining competed in year two.

The specific metrics that you use will depend on your strategic objectives and associated strategic plans. As a safety professional, you should already be familiar with various leading and lagging indicators that might apply. Here are some basic examples for the other three Strategic Areas of Safety:

- *Compliance:* A lagging indicator might be the overall percent compliance, or scores from internal compliance audits. Leading indicators could be the number of safety inspections, observations, or other similar compliance tasks performed.

- *Competence:* A lagging indicator might be one associated with a periodic competency assessment, such as the level or percentage of targeted competencies achieved. Leading indicators could be the number of trained employees, the percentage of required safety training completed, or better yet, the percentage of on-time training completion.

- *Culture:* A lagging indicator might be overall maturity level, or the percent safe score of a Safety Perception Survey. Leading indicators could be those associated with culture improvement projects, or other activities designed to improve the safety culture. Employee involvement would be beneficial, which can be measured in various ways.

Table 40 shows the performance measures for the first nine objectives in table 38. The assumption is that the lowest ranking objective was not selected for the final strategic plan, and that the organization has the resources to support the remaining nine objectives.

Table 40.   Objective Performance Measures

Key	Strategic Area	Objective Statement	Key Results
T2	Performance	Create a workplace and mindset that keeps hands out of the line of fire.	Year-on-Year Reduction in Recordable Hand Injury Rate, 40% in Year 1 and 20% in Subsequent Years
T3	Performance	Significantly reduce musculoskeletal injuries.	Year-on-Year Reduction in Recordable Musculoskeletal Injury Rate, 25% Each Year

A4	Compliance	Eliminate OSHA citations and fines.	Zero OSHA Citations Annually
O2	Culture	Create a caring and collaborative safety culture.	10% Improvement in Safety Perception Percent Safe and Perception Gap Scores
P2	Compliance	Improve the Compliance Audit Program.	20% Improvement in Average Annual Audit Score, beginning with Year 3 Audits
T6	Compliance	Exceed all client EH&S expectations.	At Least an Average Customer Survey Safety Score of 4.5 of 5.0 Within Three Years
A3	Culture	Implement quarterly safety awareness campaigns.	All four quarterly campaigns launched and tracked for implementation across the company.
W1	Competence	Develop safety leaders within the organization.	60% Improvement in Manager and Supervisor Performance Appraisal Safety Leadership Ratings by End of Year 4
A1	Compliance	Improve the Compliance Training Program.	At Least 90% On-Time Completion pf Required Training Annually, and 95% Passing Rate on First Exams

Strategic Plans

Having completed the objectives selection process, it is time to develop strategic plans for each of your objectives. Those plans should include strategic, operational, and tactical plans. Here, you should find yourself in familiar territory. Most safety professionals are involved in some form of project planning containing various levels of plans.

Developing, managing, and tracking strategic plans is a project planning process, making project planning software a great tool for it. The task hierarchies in that software are well-suited for the three levels of plans. Strategic plans represent the summary tasks, operational plans represent subtasks, and tactical plans represent sub-subtasks.

Using the example of hand injury reduction objective T2, below is a list of example strategic plans for achieving that objective.

Example Strategic Plan

Strategic Objective: Create a work environment and mindset that keeps hands out of the line of fire (objective), as measured by a 40% reduction in recordable hand injuries in year one and 20% in subsequent years (key result). To achieve this, we will [strategic plans]:

- Develop policies, programs, and procedures for the safe use of hand and power tools.

- Train employees in the safe use of hand and power tools and associated safety devices.

- Develop and implement a list of approved cutting tools that identifies the safest tool for each cutting task and the type of glove required.

- Provide safety knives for all knife-related cutting tasks to prevent contact with the cutting blade.

- Provide hands-free devices when hammers and striking implements are used, to remove hands from the line of fire.

- Institute a risk assessment and approval process for when safety knives or hands-free devices are not workable for the task.

Taking one of the strategic plans from the list above, here are example operational and tactical plans that might be developed to execute it:

Strategic Plan: Develop and implement a list of approved cutting tools that identifies the safest tool for each cutting task and the type of glove required.

Operational Plan 1: Develop a cutting task survey form for employees to help identify all cutting tasks, and complete the survey process during the first month of year one.

> *Tactical Plan:* Director of EH&S to develop the cutting task survey and oversee the survey process.

> *Tactical Plan:* Shop and Field Safety Managers to facilitate the cutting task survey in their respective shops and regions.

Operational Plan 2: Develop a List of Approved Cutting Tools based on the tasks identified in the survey during the second month of year one.

> *Tactical Plan:* Shop and Field Safety Managers to identify the safest tool and required gloves for each cutting task.

> *Tactical Plan:* Purchase and provide the required cutting tools and gloves.

Operational Plan 3: Provide training on and implement the List of Approved Cutting Tools by the end of Q1 in year one.

> *Tactical Plan:* Field Safety Managers will train on it and implement it in their respective regions.

> *Tactical Plan:* Shop Safety Managers will train on it and implement it in their respective shops.

The comprehensive, systematic, and objective strategic planning process described above offers depth and objectivity to the process that other methods do not. Whatever method you use to determine which strategic objectives to include in your strategic plan, make sure that you base your decision on objective data and sound reasoning.

To execute your strategy, you need to assign person(s) responsible for implementing the various plans. You also need to provide a timeline, including target dates for completing the respective operational and tactical plans. Again, project management software is ideal for managing those aspects of implementing the plans and allocating your available resources.

Now that you have a strategic plan, one tool that will help you communicate your plan to your organization, as well as to external stakeholders, is a written Strategic Safety (or EH&S) Plan. Since you may need to communicate that plan to external stakeholders, it should be in a format suitable for posting on the company's website and/or publishing in its Corporate Sustainability Report (CSR).

The objective of your written plan is not only to educate the organization on it, but also to motivate them to participate in its execution. Unlike the strategies of war, you want your employees to see not only the tactics used to achieve your objectives, but also the strategy that will ensure victory.

Strategy Maps

You should also develop a way to communicate your strategy and strategic objectives in an easily recognizable visual manner. An excellent tool for communicating it this way is a Strategy Map. Strategy maps are a strategic planning tool created by the Balanced Scorecard movement.

In an online article titled "Balanced Scorecard: The Comprehensive Guide," Joseph Lucco, Director of Customer Success at ClearPoint Strategies, describes a strategy map:

> One of the more powerful elements in the Balanced Scorecard framework, a strategy map, clearly visualizes your strategy for everyone to see. A strategy map is a one-page graphic that lays out your strategic objectives for you to easily communicate vision to your team.[8]

Figure 17 shows a simple strategy map, based on the four Strategic Areas of Safety, providing you with an example of how a map can visualize your overall strategy. In the example, oval shapes designate the nine strategic objectives selected above. The arrowed lines are used to show relationships between the various objectives.

Notice how the objectives under the competence strategic area support those under the performance and compliance areas. The arrangement of the strategic areas, with performance on top and culture on bottom, supports those relationships in an upward pattern of influence. Culture objectives influence objectives in the other three strategic areas.

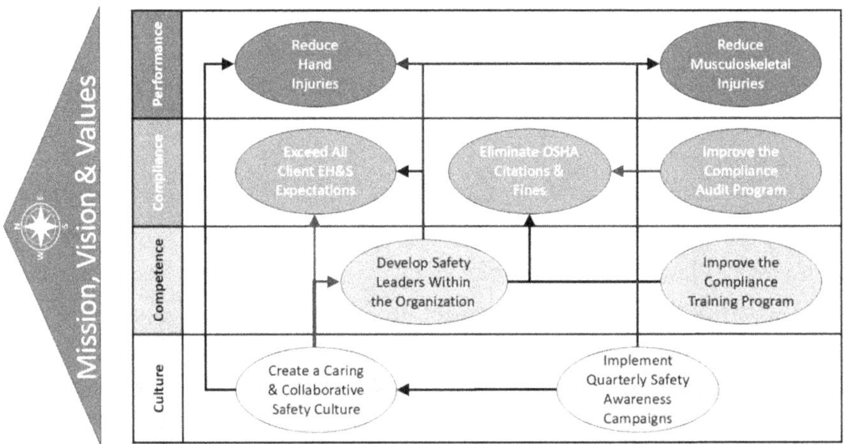

Figure 17.   Strategy Map

Strategy maps can be much more detailed than this example. More detailed examples can be found online. How detailed you make your map is a matter of preference. The key here is to visualize your strategy in such a way that the entire organization can easily understand it, and see how the various objectives contribute to the overall strategy.

The strategy map is not a map of all your strategic, operational, and tactical plans. Those should be spelled out in your written Strategic Safety Plan.

If you do not adopt a formal Balanced Scorecard approach, you can still benefit from using a strategy map as a graphic model to represent your strategic objectives. In a *Harvard Business Review* article titled, "Having Trouble with Your Strategy? Then Map It," authors Robert S. Kaplan and David P. Norton, state:

Organizations need tools for communicating both their strategy and the processes and systems that will help them implement that strategy. Strategy maps provide such a tool. They give employees a clear line of sight into how their jobs are linked to the overall objectives of the organization, enabling them to work in a coordinated, collaborative fashion toward the company's desired goals. The maps provide a visual representation of a company's critical objectives and the crucial relationships among them that drive organizational performance.[9]

Once your strategy is developed, a strategy map and written strategic plan will help you communicate it to your organization and other stakeholders. These two documents go hand-in-hand, giving the organization an overview of your strategic objectives, and the details of how you plan to achieve them.

This and the previous chapter described the strategic planning process. That process helps you to map out your journey from where your safety program is now to where you want it to be.

Developing your strategy is only half of the battle, though. The other half is executing it. Chapter Twelve covers the execution of your strategic plan. It addresses the important things to consider while you take your journey, and will help you bridge the infamous *strategy execution gap* from the outset.

Endnotes

1. Yigit Durdag, "How to Write Strategic Objectives: A Detailed Guide With Examples," *GrowForce*, accessed April 22, 2022, https://www.grow-force.com/strategic-objectives/.

2. Ted Jackson, "How to Create & Write Out Your Strategic Objectives," *ClearPoint Strategy*, accessed February 25, 2022, https://www.clearpointstrategy.com/how-to-write-strategic-objectives/.

3. Yigit Durdag, "How to Write Strategic Objectives: A Detailed Guide with Examples," *GrowForce*, accessed April 22, 2022, https://www.grow-force.com/strategic-objectives/.

4. Michael Porter, "What is strategy," *Harvard Business Review*, November-December 1996, 70.

5. Ted Jackson, "How to Create & Write Out Your Strategic Objectives," *ClearPoint Strategy*, accessed February 25, 2022, https://www.clearpointstrategy.com/how-to-write-strategic-objectives/.

6. David E. Downs, William L. Heim, "Planning, Developing, Managing and Tracking the Organizational Performance of EHS Initiatives and Programs: A Mini Safety MBA," Handout of seminar presented at American Society of Safety Engineer's SeminarFest, Las Vegas, Nevada, January 25-26, 2014, 18-24.

7. Robert S. Kaplan, "Conceptual Foundations of the Balanced Scorecard," *Harvard Business School*, 2010, 20-21, accessed February 20, 2022, https://www.hbs.edu/ris/Publication%20Files/10-074_0bf3c151-f82b-4592-b885-cdde7f5d97a6.pdf.

8. Joseph Lucco, "Balanced Scorecard: The Comprehensive Guide," *ClearPoint Strategies*, accessed February 28, 2022, https://www.clearpointstrategy.com/full-exhaustive-balanced-scorecard-example/.

9. Robert S. Kaplan and David P. Norton, "Having Trouble with your Strategy? Then Map It," *Harvard Business Review*, September 1, 2000, https://hbr.org/2000/09/having-trouble-with-your-strategy-then-map-it.

Chapter Twelve:
Executing a Strategy

Without strategy, execution is aimless. Without execution, strategy is useless. —
Morris Chang

As handy as the AAA TripTiks[R] were back in the day before electronic road navigation systems, they did nothing for you if you left them at home sitting on the kitchen counter. Likewise, the best strategic plan will do nothing to help you achieve your objectives, if your plan sits in a binder on a shelf collecting dust. A binder full of blank pages would serve you just as well, and save you all that time spent on strategic planning.

Morris Chang, the founder of Taiwan Semiconductor Manufacturing Company (TSMC), stated: "Without strategy, execution is aimless. Without execution, strategy is useless."[1] As previously stated, developing your strategy is only half of the battle. The other half, and the more challenging half, is executing it.

This chapter aims to help you ensure that your strategy execution goes as smoothly as possible and is successful in achieving your objectives. We will start by exploring what experts call the *strategy execution gap*.

Strategy Execution Gap

In their 1992 book, *"The Balanced Scorecard: Translating Strategy into Action,"* co-authors Robert S. Kaplan and David Norton stated that 90% of organizations fail to execute their strategies.[2] A few decades later we find that: "In 2016, according to the *Harvard Business Review*, it was estimated that 67% of well-formulated strategies failed due to poor execution."[3]

These two statistics suggest that there were some improvements in strategy execution over the decades. However, there is still a high failure rate because of poor execution.

If an organization's strategic plan, created to move it from its current state to its desired state, proves unsuccessful, a gap exists between the strategy and its implementation. The term for this is the *strategy execution gap*. Closing the gap is referred to as *bridging the gap*.

Closing the strategy execution gap is one of the greatest challenges expressed by executives regarding strategy. According to WorkBoard, developers of a leading enterprise Objectives & Key Results (OKR) solution using artificial intelligence: "70% of strategists are concerned they're not able to close the strategy execution gap."[4]

A review of online articles devoted to closing the gap reveals a multitude of reasons for this and advice on how to bridge it. They address gaps created during both the strategic planning and execution phases.

Strategic Planning Phase

Some strategy execution gaps arise before the execution phase even begins.

- *Unrealistic Goals:* One reason that strategy execution fails is having goals or targets that are too ambitious. Using the objective of reducing hand injuries as an example, shooting for a 75% reduction in recordable injury rates year-on-year is unreasonable. Unrealistic targets are demotivating to the organization because they are unachievable.

- *No Actionable Plans:* Another reason that strategy execution fails is not having actionable plans. Strategic objectives alone are not actionable. Strategic, operational, and tactical plans are actionable.

- *Insufficient Resources:* Being too ambitious when developing your strategic plans results in not having the resources to implement all of them. This throws a monkey wrench into the resource allocation process.

You can bridge the strategy execution gap by aiming for realistic targets, having well thought out actional plans, and scaling your plans to match available resources. Do not bite off more than the organization can chew. Consider your resources throughout the strategic planning phase.

Strategy Execution Phase

Other strategy execution gaps arise during the execution phase.

- *Poor Leadership:* Poor leadership yields poor strategy execution. Powerful leaders are passionate about strategy and will commit to the strategy. Powerful leaders inspire and motivate their teams and ensure that everyone remains focused on executing the strategy.

- *Poor Communication:* When all levels of the organization are not aware of or understand the strategy, or the role they play in it, your execution efforts are bound to fail. Employees cannot meet expectations that are not shared with them. Gaps in performance cannot be closed if not identified and communicated in a timely manner.

- *Lack of Accountability:* This applies to all levels of the organization. Executive leadership has to hold itself accountable for the success or failure of the overall execution. Managers and supervisors, likewise, have to be held accountable for the performance of their departments or teams. Individual employees have to be held accountable for fulfilling their assigned responsibilities.

You can also bridge the strategy execution gap by having strong leadership, communicating the strategy throughout the organization, and fostering a culture of accountability.

In an online article titled "Fascinating statistics on why corporate strategy breaks down," ArchPoint Consulting provides four primary reasons strategy execution breaks down, pitfalls you should avoid. The article begins with:

> Strategy execution. That wonderful, elusive thing leaders around the world covet. What a dream to develop a corporate strategy, institute it with discipline and clear communication and watch our teams adapt, allocate resources, track progress and deliver results. Unfortunately, only 33% of leaders reach this organizational nirvana, with the latest estimate that 67% of strategies fail because of poor execution.[5]

The article describes the four primary reasons why execution fails. It also provides statistics for each reason using their own research studies, and those studies provided in the best-selling book *"The 4 Disciplines of Execution: Achieving Your Wildly Important Goals."*

Pitfall 1: Employees don't understand the strategy.

> Only 5% of employees are aware of and/or understand their company's strategy.

> A mere 27% of employees and 42% of managers have access to their company's strategic plan.[6]

Pitfall 2: The organization does not know how to execute the strategy.

> Just 20% of managers think their organizations do well in allocating people across business units to support strategic initiatives.

> Over half (61%) of senior executives acknowledge their organizations do a poor job bridging the gap between strategy formulation and day-to-day implementation.[7]

Pitfall 3: Organizations don't track progress.

> In companies that consistently meet their strategic plans and goals, leadership teams meet once a month for 4-8 hours.

> A majority (92%) of organizations reported that they do not track the key performance indicators to tell them how well they are doing in competition.[8]

Pitfall 4: People aren't held accountable for performing strategic activity.

> 70% of middle managers and more than 90% of front-line employees have compensation that is not linked to the corporate strategy.

> Two-thirds of managers say that past performance is the biggest factor when making a promotion decision—but a culture that promotes strategy execution must reward teamwork, ambition, agility and a willingness to change.[9]

Based on these primary reasons businesses fail in their strategy execution efforts, we can develop five key elements of a strategy execution plan that will help ensure success in your safety strategy execution efforts.

Execution Strategy Elements

Strategy execution, like strategic planning, is a process. It takes your strategic plan and turns it into the actions needed to achieve your objectives. It is the oversight, management, and implementation of strategic, operational, and tactical plans.

Effective execution requires a structured approach. In short, your strategy execution requires an execution strategy. That strategy should include key elements proven to ensure success. These include *strong leadership, responsibility and accountability, clear communications, performance measurement,* and *adaptability and change.*

Strong Leadership

Strategy Execution Team

Strong leadership is essential for closing the strategy execution gap. One way to ensure that you have strong leadership in place is to establish a strategy execution team that plans and oversees the execution process. The makeup of the team should differ from the strategic planning team because it demands distinct qualities and skills, although at least one member from the planning team should be on it who has familiarity with the plan.

The team should be comprised of key decision makers and include powerful leaders with authority to allocate resources, assign responsibilities, ensure accountability, and drive continual improvement. If your company has a Chief Strategy Officer (CSO), that person would be an obvious choice, presuming they posess good strategy execution skills. Also, when forming your team, select members who are passionate about safety.

The ideal makeup of the team should be based on the size, structure, and nature of the organization. You should limit the strategy execution team, however, at least the core team. Randall Rollinson, President of LBL Strategies, a strategic planning and management consulting services company, stated: "In most cases we work with core planning teams of 7 to 12 with extended planning team members of 25-100... again it depends."[10]

Other experts say that your team should range from three to eight members, with the sweet spot being five. Most agree that teams larger than twelve members can get bogged down in details and delay decision making.

At a minimum, the strategy execution team must understand how to properly execute a strategy and avoid execution pitfalls. This is especially true of ArchPoint Consulting's pitfall two. Leaders need to ensure that the whole organization *does* know how to execute the strategy.

Other Strategy Leaders

Other leaders in the organization should possess strong leadership qualities, as well as strategy execution skills. This includes mid-level managers and safety function leads. They play a critical role in implementing the strategic, operational, and tactical plans associated with a safety strategy. It is likely that they represent some of extended planning team members mentioned by the President of LBL Strategies. Regarding safety function leads, there is a reason I included Safety Strategy as the first competency under Visible Leadership in the Safety Management Framework shown in table 24 in Chapter Eight.

Responsibility & Accountability

Another key element of an execution strategy is accountability. This includes establishing clear lines of responsibility for action plans, and holding individuals accountable for completing the actions assigned to them, on time and with the desired results.

An effective means of ensuring accountability is including responsibilities associated with the strategic plan as part of employee performance appraisals. Many organizations already include a safety component in appraisals, which can and should include a line item for strategic safety plan contributions.

One important thing to keep in mind during strategy execution is changes in the organization. Extended absences, employee turnover, reorganizations, and downsizing can negatively affect the success of your strategy execution if you do not reassign action plans accordingly. These potential changes should be part of your strategy review processes to identify and address them in a timely manner.

As pitfall four shows, organizations that do not hold employees accountable for performing their assigned responsibilities are likely to fall short of achieving their strategic objectives. Your plans will not come together if you plan your work but don't work your plan. It's as simple as that. To achieve their objectives, organizations need to hold employees accountable for performing their assigned responsibilities.

Clear Communications

Perhaps the most important role of the strategy execution team is ensuring that the entire organization is aware of and understands your strategy. The first pitfall listed was employees not understanding the strategy. Only 5% of employees were aware of and/or understood their company's strategy. This highlights the importance of a strategy map and written strategic plan.

Effective communication aligns the organization—top to bottom and across all functions—toward achieving the objectives. Everyone in the organization needs to know the role they play in achieving them.

Planview® is a leading platform for Strategic Portfolio Management (SPM). In an online article titled "Understanding the Importance of Strategic Alignment," they describe *strategic alignment* in the following way:

Strategic alignment is the process of working with various teams and individuals to connect their efforts to the organization's overall goals. To be effective, alignment must begin at the top with senior leaders who share the enterprise strategy throughout the organization.

In high-performing organizations, every team member has:

1. Understanding of the overall enterprise strategy,

2. Knowledge of where they fit into that strategy, and

3. Understanding of why their work matters in the bigger strategic picture.[11]

When every part of the organization is aligned, when everyone is singing from the same song sheet, the chance of successfully executing your strategy increases exponentially. Developing the best strategic plan does little good when you don't communicate it throughout the organization.

Chapter Eleven introduced strategy maps as a useful tool for communicating your overall strategy to the organization. It also recommended a detailed written strategic safety plan as a useful tool for communicating the details of your strategy. Are those enough to communicate your strategy? No! It requires a much more concerted effort.

Strategy experts recommend creating a Communications Strategy Plan to ensure your communication efforts are as effective as possible. In an online Forbes article titled "Five Components of a Successful Strategic Communications Plan," Haseeb Tariq describes a communications strategy:

A communications strategy is a plan for communicating with your target audience. It includes who you are talking to, why you are talking to them, how and when you will talk to them, what form of communication the content should take and what channels you should use to share it.[12]

Tariq described the five components listed above and concluded the article by stating the following "Bottom Line:"

A strategic communications plan can help you communicate your message to the right people at the most opportune time. By considering these five components, you can put together a solid [communication] strategy that could drive more success for your business and bring about your desired results in less time.[13]

A well-designed strategic communications plan communicates your strategy to the right people, at the right time, and in the right way. Given the critical role that communication plays during strategy execution, it behooves you to further your knowledge of effective organizational communications. For starters, below are five primary types of communication to consider including in your communications plan.

1. *Initial:* Initial communications used to launch your strategy execution process are crucial for informing internal and external stakeholders of your strategic plan, aligning the organization to achieving its objectives, and motivating employee engagement. In order to reach new employees who miss the launch, you should also incorporate your strategy into your onboarding process.

2. *Marketing:* Ongoing communications related to strategy execution should resemble a marketing campaign that promotes your strategic plan. The purpose is to sell your strategy to the stakeholders, especially internal ones, and motivate them to take action. Sharing marketing-style messages helps to keep your strategic plan at the forefront of stakeholders' minds. Various communication channels can deliver marketing-style messages, as discussed below.

3. *Updates:* Occassional communications related to strategy execution should keep internal stakeholders informed of the organization's progress in executing the strategic plans and in achieving its objectives. They should celebrate successes along the way, and drive improvements when execution efforts fall below expectations. Various communication channels can also provide updates.

4. *Feedback:* Seek and consider feedback from stakeholders throughout strategy execution. Communicate back to the stakeholders after implementing their suggestions. This two-way communication helps to promote more stakeholder engagement in the execution process.

5. *Reviews:* Periodic reviews help track progress, keep strategy execution efforts on course, and identify course corrections needed. These take the form of meetings conducted at various levels of the organization. Monthly or quarterly reviews by the strategy execution team are vitally important to track progress and make course corrections, if needed.

When implementing a strategic communications strategy, there are keys to success that will help to ensure your communications hit the mark. They involve the messages and the channels used to deliver them.

Stuart Sinclair is the founder and director of Talkfreely, an internal communication and employee experience application provider. He starts his article titled "12 Do's and Don'ts of a Great Internal Communication Strategy" with the following advice:

> What's the secret to success when it comes to internal communications? Forward planning. It's the one area of business where you simply can't get away with winging it. Unless you put some serious time and thought into your communications, you'll find your messages fail to meet their mark. Every organization needs a planned internal communication strategy to connect with its employees.[14]

The overall objectives of your strategy communications are to inform, engage, and align the organization. Presenting quality messages that are tailored to the needs of your audience helps to meet those objectives. Below are five principles to ensure effective messaging:

1. *One Message Is Not Enough:* Communicating your strategy only once is a recipe for failure. Strategy communication must be an ongoing process that uses repetition and reinforcement to ensure the message is understood, embraced, and remembered.

2. *Ensure Communications Reach Every Employee:* While the messages themselves may vary among your audiences, every subset of the organization must be included in your communications strategy. Everyone from the executive suite down to the shop floor or field can contribute to the success of your strategy execution efforts. Failing to inform, engage, and align some segments of the organization will prevent them from doing so.

3. *Don't Overcomplicate Messages:* Overly complicated messages are difficult for employees to grasp, and they may gloss over the messages when reading them. This is a case where less is more. To communicate effectively, break down your messages into a few clear and concise concepts that your audience can digest.

4. *Include a Call to Action:* Messages without a call to action are just statements and do little to foster employee engagement. What should your employees and other stakeholders do with the information you are presenting? Your audience is far more likely to contribute to the strategy execution's success if you include a call to action.

5. *Avoid Communication Overload:* Communicating your strategic plan only once is not enough. Overdoing it, however, can also be detrimental to your strategy execution efforts. Excessive communication can appear manipulative, disrespectful of your employees' time, and even insulting to their intelligence. Employees may come to resent your strategy communications, or ignore them altogether.

How you distribute your messages is just as important as its content. If a well-designed strategic communications plan communicates your strategy to the right people at the right time, in the right way, what is the right way? Well, that depends on the communication type and the target audience.

There are numerous channels of communication to consider, each having their own benefits.

- *Meetings:* Real-time, two-way communication is by far the best, but not always practical. This should be your first choice for the initial rollout of your strategy, as well as reviews, either in-person or virtual, because they allow for two-way communications.

For marketing and update communications, there are a wide variety of channels available, using either print or digital media.

- *Print Media:* Hard copy memos, brochures, newsletters, mailers, posters, and reports all still have a place in a good communication strategy.
- *Digital Media:* Email, intranet sites, digital display systems, social media, and mobile platforms provide excellent ways to deliver marketing or promotional messages, as well as updates.

In an online *Forbes* article titled "7 Mistakes to Avoid When Communicating Your Strategy," contributor Sherzod Odilov concludes with an excellent summary of the importance and benefits of strategy communications. He states:

> Effective communication is the linchpin of successful strategy execution. By avoiding common mistakes and implementing a comprehensive communications plan, you can ensure that your strategy is clearly understood, embraced by employees, and achieves its intended goals.[15]

Performance Measurement

Another appalling statistic from ArchPoint Consulting studies was 92% of organizations do not track the key performance indicators to tell them how well they are doing. Imagine putting all that effort into developing and executing a strategy and not tracking your progress.

Would you go into major surgery with no monitoring equipment? Imagine not knowing your heart rate, blood pressure, respiratory rate, blood oxygen levels, etc., during or right after extensive surgery? Imagine not having any follow-up examinations or diagnostic testing. You would not know whether you were surviving through or recovering from the surgery, or if the surgery even resolved your medical issue.

Tracking every aspect of your strategy execution is critical to success. This is especially true of the percentage completion of actions and your key results. The former measures progress, while the latter measures effectiveness, both of which are measures of success.

Strategy execution requires knowing the vital statistics to ensure you are heading in the right direction, and to know if your plan is doing what it was designed to do. Tracking these vital signs on a real-time basis helps you to identify problems early on so that you can make adjustments or course corrections to keep on track and maximize effectiveness.

Continuing with the surgery and recovery analogy, who needs to have access to your vital signs? Everyone on your surgical and post-operative care team needs that information. The same holds true for your strategy execution team and everyone involved in the strategy execution process. Your performance measures need to be available to all of them.

Adaptability & Change

Adaptability is the capacity to adjust to new or changing situations. During the strategy execution phase, you will face new or changing situations. Resource changes, especially, will significantly affect the execution process.

You may recall from Chapter One that John C. Maxwell stated: "A strategy that doesn't take into account resources is doomed to failure."[16] Another important role of the strategy execution team, therefore, is allocating the resources (time, people, money) throughout the strategy execution process, especially when situations change. Even if you planned well during the planning phase, you will most likely need to reallocate resources during execution due to changing circumstances. Recall what Field Marshal Helmuth von Moltke wrote about no plan of operations extending beyond the first contact with the enemy.

A key principle of effective project and resource management is monitoring the workload and associated resources. Using project management or strategy execution software, especially if it includes task assignments, helps make allocating or reallocating resources much easier. By monitoring the associated resources, the strategy execution team can identify constraints and resolve them, or recognize underutilization and take advantage of it.

Another key principle of effective project and resource management is aligning resources with priorities. The department needs to use its limited resources in the best way possible. One method of prioritizing resources is assigning them to the action plans that have the greatest value to the strategic plan. This is where the Weighted Value Analysis, described in Chapter Eleven, comes in handy.

Another method of prioritizing resources, not yet described, is using the Eisenhower Matrix, a simple and effective time or resource management tool. It is very similar to the P.I.C.K. Matrix dexcribed in step three of the objective selection process, and is used in the same manner.

The matrix traces its roots to an address made by President Dwight D. Eisenhower. While addressing the Second Assembly of the World Council of Churches in 1954, Eisenhower quoted a college president: "This President said, 'I have two kinds of problems, the urgent and the important. The urgent are not important, and the important are never urgent.'"[17]

Eisenhower's quote was later applied to a matrix using *importance* and *urgency* to plot actions in one of four quadrants, as shown in figure 18. Stephen R. Covey popularized the concept in "The 7 Habits of Highly Effective People," with his Time Management Matrix.

The importance and urgency are rated on a scale of one to ten, the same as the P.I.C.K. Matrix scales, with one being the lowest and ten being the highest. Importance, for the sake of resource allocation, relates to the value of the action in achieving the strategic objective. Urgency relates to the scheduled times for starting or completing actions, with those coming due sooner being more urgent.

In figure 18, it is important to notice a key difference from the P.I.C.K. Matrix. The urgency scale is reversed, with the highest value to the left and not the right of the scale.

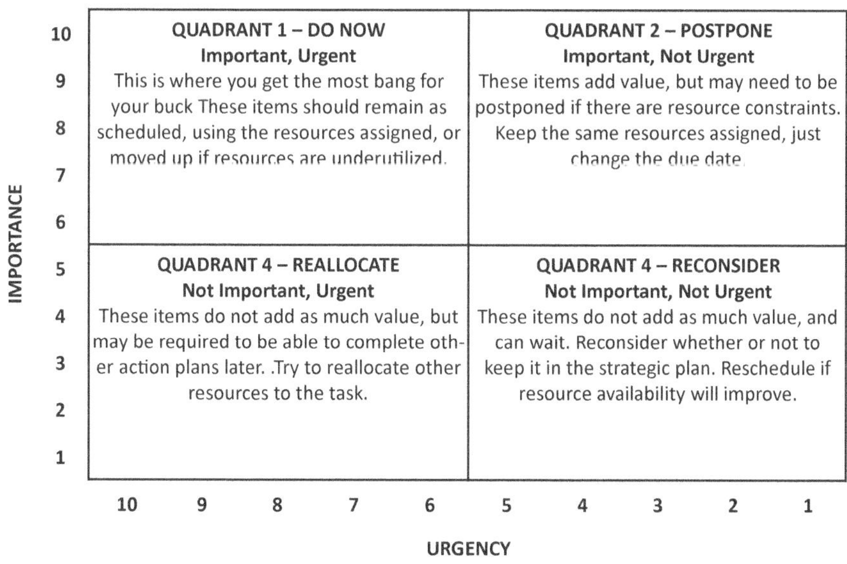

Figure 18.   Eisenhower Matrix

Below are descriptions of the four quadrants.

Quadrant 1 (Important, Urgent): Actions that should be completed as scheduled despite any resource constraints, or scheduled sooner if resources are underutilized.

Quadrant 2 (Important, Not Urgent): Actions that can be postponed if there are resource constraints. Reschedule them for when you expect the assigned resources to be available.

Quadrant 3 (Not Important, Urgent): Actions not important in and of themselves but may support subsequent action plans. If possible, allocate other resources within the organization to complete them.

Quadrant 4 (Not Important, Not Urgent): Actions you should reconsider whether to include them as part of the strategic plan. If resource constraints prohibit completing an action, eliminate it from the plan. You can always add or reschedule it later if circumstances change.

When resource availability changes, the strategic plan needs to be changed to compensate. Making those changes requires adequate monitoring of the workload and associated resources, and being able to prioritize action plans. Using a prioritization method like those described above helps to make the process more objective, and thus more effective.

Each of the elements described above plays a critical role in ensuring that you bridge the execution gap. Ignoring them will earn you a place in the strategy execution hall of shame for failing to achieve your objectives. Do not become another failure statistc.

In the introduction, I mentioned taking a data-driven strategic approach to injury reduction, and how we achieved a 77% reduction in TRIR over five years. Two important lessons I learned from that involved strong leadership and communications.

Strong Leadership: While we did not formally have a strategy execution team, we did have a senior leadership team involved in tracking execution progress and driving accountability down through the organization. The team consisted of the division president and his senior vice presidents of business segments from around the world That division consisted of around 70% of the new joint venture company, calculated on a man-hours worked basis, so it made a significant contribution to the success of the joint venture.

The president was involved in formulating our strategy, and was involved with and deeply supportive of its execution. He personally held the senior vice presidents accountable for their progress and results. That level of leadership and support was extremely effective, and cemented for me the importance of having strong leadership during strategy execution.

Communications: Every quarter we held a dedicated safety review meeting with the president and his senior vice presidents. While I presented that video conference call, which included the ongoing progress and results of our strategy, the president was directly involved with engaging his senior leadership team regarding their progress and results, whether good or bad.

Charts and data tables were provided in my presentations for each segment of the business. The senior vice presidents were able to share those with their own leadership teams, effectively communicating results down through the organization. The progress and results were also shared during monthly conference calls with our global safety management community.

Every segment of the division was able to see not only their data, but that of the other business segments. Both successes and failures were shared throughout the organization. These calls also allowed senior vice presidents to share the reasons for their successes and failures. Collectively, this created a highly effective teamwork approach to achieving our objectives.

Both the strong leadership and structured communications among the senior leadership team played a significant part in our successful strategy execution. It was a blessing to have that level of support, and the platform for focused communications, to say the least.

As I already mentioned, "Zig" Ziglar was one of my favorite motivational speakers when I was a young man. He is considered by many to be the father of motivational speaking. What drew me to Ziglar, besides his catchy professional name, was his physical speaking style, his unique cadence and voice inflection, and his concise and clear nuggets of wisdom. He had the ability to condense important principles into short, pithy sayings.

One such nugget is: "Execution is the bridge between goals and accomplishment."[18] I can hear him saying it using his characteristic drawn-out inflection on the words *execution* and *bridge* to emphasize the key point. That bridge needs to be designed and constructed to support the execution process in order for you to cross it. This chapter has provided you with the key design elements for constructing that bridge.

Endnotes

1. John Moore, "Best Quote on Strategy vs. Execution," *Brand Autopsy*, accessed June 15, 2025, https://brandautopsy.com/2010/09/best-quote-on-strategy-vs-execution-.html.

2. Kate Gibson, "5 Reasons Strategy Execution Fails," *Harvard Business School*, December 21, 2023, accessed July 11, 2025, https://online.hbs.edu/blog/post/why-do-strategic-plans-fail.

3. Ron Carucci, "Executives Fail to Execute Strategy Because They're Too Internally Focused," *Harvard Business Review*, November 13, 2017, accessed July 11, 2025, https://hbr.org/2017/11/executives-fail-to-execute-strategy-because-theyre-too-internally-focused.

4. "How Top Companies are Closing the Strategy Execution Gap," *Quantive* (now *WorkBoard*), accessed July 11, 2025, https://quantive.com/resources/articles/strategy-execution-gap.

5. ArchPoint, "Fascinating Statistics on Why Corporate Strategy Breaks Down," *ArchPoint Consulting*, June 28, 2019, accessed on July 11, 2025, https://archpointconsulting.com/strategy/fascinating-statistics-on-why-corporate-strategy-breaks-down.

6. Ibid.

7. Ibid.

8. Ibid.

9. Ibid.

10. Randall Rollinson, "What is the Optimal Approach and Team Size for a Strategic Planning Effort?" *LBL Strategies*, November 17, 2014, accessed June 17, 2025, https://www.lblstrategies.com/what-is-the-optimal-approach-and-team-size-for-a-strategic-planning-effort/.

11. "Understanding the Importance of Strategic Alignment," *Planview*, accessed July 5, 2025, https://www.planview.com/resources/articles/understanding-the-importance-of-strategic-alignment/.

12. Haseeb Tariq, "Five Components of a Successful Strategic Communications Plan," *Forbes*, June 22, 2021, accessed July 5, 2025, https://www.forbes.com/councils/forbescommunicationscouncil/2021/06/22/five-components-of-a-successful-strategic-communications-plan/.

13. Ibid.

14. Stuart Sinclair, "12 Do's and Don'ts of a Great Internal Communication Strategy," *Talkfeely*, May 9, 2023, accessed July 7, 2025, https://www.talkfreely.com/blog/internal-communication-strategy.

15. Sherzod Odilov, "7 Mistakes to Avoid When Communicating your Strategy," *Forbes*, July 21, 2024, accessed July 5, 2025, https://www.forbes.com/sites/sherzododilov/2024/07/21/7-mistakes-to-avoid-when-communicating-your-strategy/.

16. John C. Maxwell, *How Successful People Think: Change Your Thinking, Change Your Life* (New York, Center Street, 2009), 56

17. Dwight D. Eisenhower, "Address at the Second Assembly of the World Council of Churches," *The American Presidency Project*, August 19, 1954, accessed July 9, 2025, https://www.presidency.ucsb.edu/documents/address-the-second-assembly-the-world-council-churches-evanston-illinois..

18. Zig Ziglar, "50 Strategy Execution Quotes: Wisdom and Inspiration for Successful Implementation," *RoundMap*, accessed July 13, 2025, https://roundmap.com/50-strategy-execution-quotes/.

Conclusion

This book began discussing safety excellence in the introduction. If you recall, Shawn M. Galloway provided us with a different way of defining safety excellence using the following three qualities of an organization that has achieved it:

Having now read this book, can you see how the concept of strategic safety can help your organization develop those three qualities?

Looking back, the book provided processes and techniques that, when properly applied, will help you achieve and repeat great results using strategy. That satisfies the first part of Galloway's definition of safety excellence.

Through implementing those processes and applying those techniques, in both strategic planning and execution, you will develop keener insight into what led to those results. That satisfies the second part of Galloway's definition of safety excellence.

Reviewing and revising your strategic plan annually drives continual improvement. Identifying course corrections needed during strategy execution also creates a cultural mindset of continuous improvement. That satisfies the third part of Galloway's definition of safety excellence.

That brings us to a final observation. In the last paragraph of his book titled *"The Mind of the Strategist,"* Kenichi Ohmae (Mr. Strategy) wrote:

> All this leads me to a final observation. Strategic success cannot be reduced to a formula, nor can anyone become a strategic thinker merely by reading a book. Nevertheless, there are habits of mind and modes of thinking you can build through practice to help you free the creative power of your subconscious and improve your odds of coming up with winning strategic concepts. The main purpose of this book is to encourage you to do so and to point out the direction you should pursue.[2]

The same can be said about the main purpose of this book. Hopefully, it has convinced you of the value of taking a more strategic approach to safety, encouraged you to further your knowledge of the subject, and provided you with a course to take on your strategic safety journey.

The ball is in now your court! You must decide what you will do with the information presented. I suggest you start your own journey in strategic safety, one step at a time.

Lao Tzŭ, a Chinese philosopher and author, wrote the following in his book, "*Tao Te Ching*," the foundational text of Taoism: "A journey of a thousand miles starts from beneath your feet."[3] (Also translated: "A journey of a thousand miles begins with a single step.") The journey starts from where you are right now, and then proceeds one step at a time until you get where you want to be.

Taoism is a philosophical and religious tradition in China that emphasizes harmony with the Tao. The Tao refers to the "Path" or the "Way." Sun Tzŭ viewed the Tao as a strategic tool in war, and integrated Taoist philosophy throughout *The Art of War*.

You do not need to be an adherent to Taoism to appreciate having a clear path forward. That is what this book aimed to do for you, help you chart a course for achieving safety excellence.

Now here you stand, ready to start your own journey in strategic safety, if you choose to take one. If so, here is a path you can take, at whatever pace your circumstances allow, to put strategic safety into practice:

1. *Make a Commitment:* Commit to being more strategic in your approach to safety.

"Motivation is what gets you started. Commitment is what keeps you going" is a quote often attributed online to Jim Rohn, author, motivational speaker, and mentor to Tony Robbins. No definitive source for that quote could be found. Still, it offers sound wisdom in this case. While you might be motivated by this book to practice strategic safety, a personal commitment is needed for you to persevere to the point of succeeding with it.

2. *Don Your Strategic Thinking Cap:* Apply the principles of strategic thinking when addressing safety problems that arise.

Chapter Four redefined strategic thinking as: "The ability to generate insights on a continual basis to achieve safety excellence." Strategic thinking involves using strategic intuition as opposed to expert intuition. Rather than make a snap decision based on your expert intuition the next time a safety problem arises, take the time to let your strategic intuition kick in. Chapter Four also provided principles and processes to foster strategic intuition, techniques that you can employ regardless of any strategic planning efforts.

What nagging safety problem are you currently seeking a solution for? Apply the strategic thinking process to it. Recall that the process is similar to the RCA process, and includes *information gathering, information analysis, and strategy formulation.* You may just surprise yourself by coming up with a creative, imaginative solution that resoles your problem once and for all.

3. *Perform Some Strategic Analysis:* Start performing the various strategic analyses presented.

Like strategic thinking, strategic analysis can be performed regardless of any comprehensive strategic planning efforts. Perform one analysis at a time and take your time. The quality of the results is more important.

If your injury rates are higher than desired, I would recommend starting with a cascading-style Attribute Analysis of historical injuries. If you think your safety culture is poor, I recommend starting with a Maturity Assessment or Safety Perception Survey. These are the two strategic areas that I often focused on, due to their impact on overall safety performance.

Granted, the analyses described in Part III of this book take time. Set aside a portion of your time each week to devote to performing one analysis at a time. The analyses are also best performed with a small team. If that is not doable given your circumsastances, you can still perform them on your own. You will still benefit by seeing the value they offer.

4. *Start Developing a Safety Strategy:* The next time you identify a detrimental safety issue, develop a strategy to resolve it.

Your first strategy does not need to be all-encompassing; it just needs to employ some of the processes presented so that you can start learning them. The important part is experiencing the value of strategic planning. Select one of the desired end states you want to achieve and develop a plan to achieve it. Chapter Five provided an example using a reduction in hand injuries.

If you do want to start a more comprehensive safety strategy, but lack the time and resources to do it all at once, I suggest doing it one focus area at a time. Before long, you'll find yourself with a complete strategy.

You may be familiar with the old adage, "Once begun is half done." Also, the slowest progress in any race is made when coming off the starting blocks or at the starting line. Once you get up to speed, momentum starts to kick in and brings you to the finish line.

5. *Fifth: Execute Your Strategy:* With your strategic safety plan in place, develop an execution strategy and start executing your plan.

Be sure to take into consideration the strategy execution elements described in Chapter Twelve, commensurate with the scope of your strategic plan. They will help you avoid the various pitfalls of poor execution that lead to failure.

In short, become a student of strategy. Start developing and practicing your strategic thinking, analysis, and planning skills today. Like the countless military and business strategists that went before you, you'll be glad that you did.

Endnotes

1. Shawn M. Galloway, "Rethinking Safety Excellence," *Occupational Health & Safety*, October 1, 2016, accessed October 13, 2023, https://ohsonline.com/articles/2016/10/01/rethinking-safety-excellence.aspx

2. Kenichi Ohmae, *The Mind of the Strategist: The Art of Japanese Business*, (New York: McGraw-Hill,1982), 277.

3. Stephen Mitchell, *Tao Te Ching: A New English Version*, (New York: Harper & Row PS TM, 1988), Poem 64, accessed July 16, 2025, https://cpb-us-w2.wpmucdn.com/u.osu.edu/dist/5/25851/files/2016/02/taoteching-Stephen-Mitchell-translation-v9deoq.pdf

Index

About The Author

Don Fritz is a Certified Safety Professional (CSP) who has spent over 25 years improving safety performance in heavy industries, including independent service providers for the power generation, oil and gas, industrial, and aerospace industries. He worked in Environmental, Health & Safety (EH&S) roles at all levels of the organization, including the business unit, divisional, and corporate levels. His experience includes safety oversight for both domestic and international businesses, and he prides himself on having repeatedly achieved annual double-digit injury rate reductions. He also has a passion for helping others in the EH&S field develop their professional careers.

Mr. Fritz began his career in the U.S. Navy's Nuclear Power Program. Upon graduating from the prestigious Navy Nuclear Power School, he was picked up as a junior staff instructor. He later served aboard a ballistic missile submarine. That experience served as a springboard to his long and fulfilling career in the power generation industry, starting in operations, then transitioning to Environmental, Health & Safety.

Mr. Fritz is now the Principal of Sanovus LLC, a consulting and professional development company. Visit his website at www.sanovus.com to find free and low cost resources, including some that were presented in this book.